# APPLICATIONS OF SOCIAL NETWORK ANALYSIS FOR BUILDING COMMUNITY DISASTER RESILIENCE

## WORKSHOP SUMMARY

Sammantha L. Magsino, *Rapporteur*

Board on Earth Sciences and Resources
Division on Earth and Life Studies

NATIONAL RESEARCH COUNCIL
*OF THE NATIONAL ACADEMIES*

THE NATIONAL ACADEMIES PRESS
Washington, D.C.
**www.nap.edu**

**THE NATIONAL ACADEMIES PRESS • 500 Fifth Street, N.W. • Washington, DC 20001**

**NOTICE:** The project that is the subject of this report was approved by the Governing Board of the National Research Council, whose members are drawn from the councils of the National Academy of Sciences, the National Academy of Engineering, and the Institute of Medicine. The members of the committee responsible for the report were chosen for their special competences and with regard for appropriate balance.

This study was supported by the U.S. Department of Homeland Security under Award No. HSHQDC-08-C-00176. Any opinions, findings, and conclusions or recommendations contained in this document are those of the authors and do not necessarily reflect the views of the agencies that provided support for the project. Mention of trade names, commercial products, or organizations does not constitute their endorsement by the sponsoring agencies.

International Standard Book Number-13: 978-0-309-14094-2
International Standard Book Number-10: 0-309-14094-3

Additional copies of this report are available from the National Academies Press, 500 Fifth Street, N.W., Lockbox 285, Washington, DC 20055; (800) 624-6242 or (202) 334-3313 (in the Washington metropolitan area); Internet http://www.nap.edu

*Cover:* Social network image courtesy of Carl Latkin.

Copyright 2009 by the National Academy of Sciences. All rights reserved.

Printed in the United States of America

THE NATIONAL ACADEMIES

*Advisers to the Nation on Science, Engineering, and Medicine*

The **National Academy of Sciences** is a private, nonprofit, self-perpetuating society of distinguished scholars engaged in scientific and engineering research, dedicated to the furtherance of science and technology and to their use for the general welfare. Upon the authority of the charter granted to it by the Congress in 1863, the Academy has a mandate that requires it to advise the federal government on scientific and technical matters. Dr. Ralph J. Cicerone is president of the National Academy of Sciences.

The **National Academy of Engineering** was established in 1964, under the charter of the National Academy of Sciences, as a parallel organization of outstanding engineers. It is autonomous in its administration and in the selection of its members, sharing with the National Academy of Sciences the responsibility for advising the federal government. The National Academy of Engineering also sponsors engineering programs aimed at meeting national needs, encourages education and research, and recognizes the superior achievement of engineers. Dr. Charles M. Vest is president of the National Academy of Engineering.

The **Institute of Medicine** was established in 1970 by the National Academy of Sciences to secure the services of eminent members of appropriate professions in the examination of policy matters pertaining to the health of the public. The Institute acts under the responsibility given to the National Academy of Sciences by its congressional charter to be an adviser to the federal government and, upon its own initiative, to identify issues of medical care, research, and education. Dr. Harvey V. Fineberg is president of the Institute of Medicine.

The **National Research Council** was organized by the National Academy of Sciences in 1916 to associate the broad community of science and technology with the Academy's purposes of furthering knowledge and advising the federal government. Functioning in accordance with general policies determined by the Academy, the Council has become the principal operating agency of both the National Academy of Sciences and the National Academy of Engineering in providing services to the government, the public, and the scientific and engineering communities. The Council is administered jointly by both Academies and the Institute of Medicine. Dr. Ralph, J. Cicerone and Dr. Charles M. Vest are chair and vice chair, respectively, of the National Research Council.

**www.national-academies.org**

## PLANNING COMMITTEE FOR THE WORKSHOP ON APPLICATIONS OF SOCIAL NETWORK ANALYSIS FOR BUILDING COMMUNITY DISASTER RESILIENCE

SUSAN L. CUTTER, *Chair*, University of South Carolina
KATHLEEN M. CARLEY, Carnegie Mellon University
WILLIAM A. V. CLARK, University of California, Los Angeles
ERIC HOLDEMAN, ICF International
RANDOLPH H. ROWEL, Morgan State University
MONICA SCHOCH-SPANA, University of Pittsburgh Medical Center

*National Research Council Staff*

SAMMANTHA MAGSINO, Study Director
JARED P. ENO, Research Associate
ERIC J. EDKIN, Program Assistant

## GEOGRAPHICAL SCIENCES COMMITTEE

WILLIAM L. GRAF, *Chair*, University of South Carolina
LUC E. ANELIN, Arizona State University
WILLIAM A.V. CLARK, University of California
CAROL P. HARDEN, University of Tennessee
CALESTOUS JUMA, Harvard University
JOHN A. KELMELIS, The Pennsylvania State University
VICTORIA A. LAWSON, University of Washington
SUSANNE C. MOSER, Susanne Moser Research & Consulting
THOMAS M. PARRIS, ISciences LLC
NORBERT P. PSUTY, Rutgers University
DAVID R. RAIN, The George Washington University

*National Research Council Staff*

CAETLIN OFIESH, Associate Program Officer
JARED P. ENO, Research Associate
TONYA FONG YEE, Senior Program Assistant

## BOARD ON EARTH SCIENCES AND RESOURCES

GEORGE M. HORNBERGER, *Chair*, University of Virginia, Charlottesville
KEITH C. CLARKE, University of California, Santa Barbara
DAVID J. COWEN, University of South Carolina
WILLIAM E. DIETRICH, University of California, Berkeley
ROGER M. DOWNS, The Pennsylvania State University, University Park
JEFF DOZIER, University of California, Santa Barbara
KATHERINE H. FREEMAN, The Pennsylvania State University, University Park
WILLIAM L. GRAF, University of South Carolina
RUSSELL J. HEMLEY, Carnegie Institution of Washington, Washington, D.C.
MURRAY W. HITZMAN, Colorado School of Mines, Golden
EDWARD KAVAZANJIAN JR., Ira A. Fulton School of Engineering
LOUISE H. KELLOGG, University of California, Davis
ROBERT B. McMASTER, University of Minnesota
CLAUDIA INÉS MORA, Los Alamos National Laboratory
BRIJ M. MOUDGIL, University of Florida
CLAYTON R. NICHOLS, Idaho National Engineering and Environmental Laboratory (retired), Standpoint
JOAQUIN RUIZ, University of Arizona, Tucson
WILLIAM W. SHILTS, Illinois State Geological Survey, Champaign
RUSSELL STANDS-OVER-BULL, BP American Production Company, Pryor, Montana
TERRY C. WALLACE JR., Los Alamos National Laboratory, New Mexico
HERMAN B. ZIMMERMAN, National Science Foundation (Retired)

*National Research Council Staff*

ANTHONY R. DE SOUZA, Director
ELIZABETH A. EIDE, Senior Program Officer
DAVID A. FEARY, Senior Program Officer
ANNE M. LINN, Senior Program Officer
SAMMANTHA L. MAGSINO, Program Officer
CAETLIN M. OFIESH, Associate Program Officer
JENNIFER T. ESTEP, Financial Associate
JARED P. ENO, Research Associate
NICHOLAS D. ROGERS, Research Associate
COURTNEY R. GIBBS, Program Associate
TONYA E. FONG YEE, Senior Program Assistant
ERIC J. EDKIN, Program Assistant

# Acknowledgments

In response to a request by the Department of Homeland Security, the National Research Council formed an ad hoc committee to organize a two-day workshop to discuss the use of social network analysis (SNA) for the purpose of building community disaster resilience. The workshop was held February 11-12, 2009, in Washington, D.C., and engaged a group of approximately 30 researchers in the fields of SNA and resilience science as well as emergency management practitioners from different regions of the country. Gaps in knowledge regarding SNA and its use for constructing designed networks for the purpose of increasing resilience were discussed, as were areas of research that could fill those gaps.

The National Research Council greatly acknowledges the work of the planning committee that designed this workshop. Susan L. Cutter of the University of South Carolina was the workshop moderator and served as chair of the workshop planning committee. Kathleen Carley, Carnegie Mellon University; William A.V. Clark, University of California, Los Angeles; Eric Holdeman, ICF International; Randolph H. Rowel, Morgan State University; and Monica Schoch-Spana, University of Pittsburgh Medical Center served as members of the planning committee. All put a great deal of time, thought, and effort into planning an agenda, identifying and inviting speakers and attendees, and preparing a detailed agenda book that included a list of select references on SNA and resilience. Committee members also served as moderators and rapporteurs for individual breakout sessions.

The National Research Council would like to thank Kathleen Carley; Carnegie Mellon University; Fran H. Norris, Dartmouth Medical School; Carl Latkin, Johns Hopkins Bloomberg School of Public Health; and Michael Byrne, ICF International for providing excellent workshop presentations intended to orient attendees regarding the subject matter to be discussed. Additionally, the workshop would not have been successful without the important contributions of those who attended the event. A complete list of participants is found in Appendix C of this document. Discussions were informative, professional, and conducted in a cooperative spirit among, in large part, individuals who do not often have the opportunity to collaborate.

This workshop summary was prepared by National Research Council staff following the workshop. It represents the findings of the workshop participants as interpreted by a rapporteur.

This workshop summary has been reviewed in draft form by persons chosen for their diverse perspectives and technical expertise in accordance with procedures approved by the National Research Council's Report Review Committee. The purposes of this review are to provide candid and critical comments that will assist the institution in making the published summary as sound as possible and to ensure that the summary meets institutional standards of objectivity, evidence, and responsiveness to the study charge. The review comments and draft manuscript remain confidential to protect the integrity of the deliberative process. We wish to thank the following for their participation in the review of this summary:

William A.V. Clark, University of California, Los Angeles
Darrell Darnell, District of Columbia Homeland Security and Management Agency
Jeffrey Johnson, East Carolina University
Naim Kapucu, University of Central Florida
Ann Patton, Ann Patton Company LLC
Monica Schoch-Spana, University of Pittsburgh Medical Center.

Although the reviewers listed above have provided many constructive comments and suggestions, they were not asked to endorse, nor did they see, the final draft of the workshop summary before its release. The review of this summary was overseen by the Division on Earth and Life Studies. The division was responsible for making certain that an independent examination of this summary was carried out in accordance with institutional procedures and that all review comments were carefully considered. Responsibility for the final content of this summary rests entirely with the author and the National Research Council.

# Contents

# Summary

## PURPOSE OF THE WORKSHOP

Social Network Analysis (SNA) is the identification of the relationships and attributes of members, key actors, and groups that social networks comprise. The National Research Council (NRC), at the request of the Department of Homeland Security (DHS), formed an ad hoc committee to plan a two-day workshop on the use of SNA for the purpose of building community disaster resilience. The workshop, held February 11-12, 2009, was designed to provide guidance to the DHS on a potential research agenda that would increase the effectiveness of SNA for improving community disaster resilience. Explored were the state of the art in SNA and its applications in the identification, construction, and strengthening of networks within U.S. communities. Workshop participants discussed current work in SNA focused on characterizing networks; the theories, principles and research applicable to the design or strengthening of networks; the gaps in knowledge that prevent the application of SNA to the construction of networks; and research areas that could fill those gaps. Elements of a research agenda to support the design, development, and implementation of social networks for the specific purpose of strengthening community resilience against natural and human-made disasters would be discussed. Box S-1 provides definitions of terms commonly used during the workshop.

## WORKSHOP PLANNING

A planning committee designed the workshop to explore how SNA could be applied during all phases of the disaster cycle. The planning committee invited researchers with expertise in resilience science and in SNA for a variety of applications (e.g., anti-terrorism and public health) to participate in the workshop and discuss the states of the art and science in their respective fields. Emergency management practitioners with experience responding to disasters were invited so that the needs of community leaders with their "boots on the ground" could be considered. The committee included participants from different geographical regions and with varying disaster experiences so that a broad range of issues and perspectives could be explored.

Sessions of the workshop were devoted to specific themes. In the context of disaster preparedness, the roles of SNA and communication in enhancing the functional, structural, and interactional connections between networks were discussed. Barriers to the use of SNA for planning activities that decrease the impact of disasters (e.g., interventions) were also discussed. Workshop participants considered how SNA could be used to make network ties between organizations more productive, and how SNA could be applied during and following a disaster to make improvisational responses—those planned once needs and resources are identified—more flexible. How individuals and communities could be engaged to promote collective behavior when preparing for, responding to, and recovering from disasters was considered.

**BOX S-1**
**Definitions of Key Workshop Terms**

The following are definitions of key terms used in the study of social networks, social network analysis, resiliency science, and research translation used during this workshop.

**Resilience.** The response to stress at individual, institutional, and societal levels categorized as the characteristics that promote successful adaptation to adversity.

**Social network.** The interactions between people and organizations, including who knows, works with, or communicates with whom, that can be mapped. The data and information found on tools such as Facebook and the Enron Email Corpus are examples of social networks.

**Social network analysis.** The process of analyzing a social network and identifying key actors, groups, vulnerabilities, and redundancies as well as the changes in these variables.

**Social networking.** The process of creating, maintaining, or altering one's network and to one's advantage by using the network to gain resources or influence, or to mobilize activity.

**Social network analysis tools.** The set of tools, technologies, metrics, models, and visualization techniques used for social network analysis. These may include data extraction tools, link analysis, statistical techniques, and graph theory techniques using programs such as AutoMap, ORA, UCINET, and Pajek.

**Social network theory.** The set of theories for forecasting, reasoning about, and understanding how social networks form, are maintained, and evolve, and the role of variables such as social networking tools, media, and stress in affecting the emergence, utilization, management, and change in social networks.

**Social network tools.** A set of computational techniques that enable individuals and groups to engage in social networking by monitoring and interacting within the networks with which they are connected. Facebook, MySpace, and Twitter are examples of social networking tools.

**Translation research.** The research aimed at enhancing the movement of research results from the scientific to the applied realms.

## WORKSHOP SUMMARY

This document summarizes the major points and ideas expressed during the workshop as documented by the rapporteur. As such, the summary reflects the specific topics emphasized by workshop presentations and discussions and may not be a comprehensive summary of all relevant topics and issues. Viewpoints expressed in this summary do not necessarily represent the views of the workshop planning committee or the NRC, nor does the summary contain conclusions or recommendations.

## GENERAL WORKSHOP THEMES

A robust scientific literature on SNA exists, and literature in disaster and community resilience is emergent. However, workshop participants noted that disjunctions exist between SNA theory and its application, and between the SNA research and emergency management communities. Workshop participants discussed how properly targeted research in networking theory, the social and resiliency sciences, and research translation, conducted in parallel with the development of SNA tools designed specifically for and with emergency management practitioners, could facilitate the adoption of SNA by the emergency management community. The adoption of SNA has the potential to revolutionize the way organizations and communities function in general, and prepare and respond to disasters in specific.

SNA allows study of complex human systems through the visualization and characterization of relationships between people, groups, and organizations. A graphical representation of a social network that shows individual network members (defined as nodes) and their linkages (defined as ties) could be a product of the analysis (see Figure S-1). The impact of information or activities on individuals and the network as a whole may be analyzed and predicted for different scenarios and options. Because SNA can reveal the characteristics, composition, and structure of networks at a given time and over time, SNA could be an important tool for understanding how parts of the community work or could work together to plan for and respond to disasters. SNA has been used to inform policy in areas such as terrorism prevention and public health improvement, and could facilitate decision making related to the improvement of community disaster resilience.

Community resilience, in sociological terms, is the ability of a community or social unit to withstand external shocks, such as disasters, to its infrastructure. Community resilience emerges from a community's ability to adapt to stress and return to healthy functioning. The speed with which a community can mobilize and use resources during and following a disaster is strongly dependent on its abilities to adapt to change. The strength of its social networks is a factor. Building community resilience is a process that develops the capacities that allow communities to adapt. The building of disaster resilience can be considered a strategy for disaster readiness. Incremental improvements in resilience can significantly improve the capacity of a community to prepare for, respond to, and recover from disasters. However, just as a community may change with time, a community's response to a disaster may change with time. A disaster that has little impact on a community at one time may have a devastating impact on it at another time. An understanding of the dynamic nature of resilience is essential for good planning. Successful building of resilience is dependent on the reduction of risk to individuals and

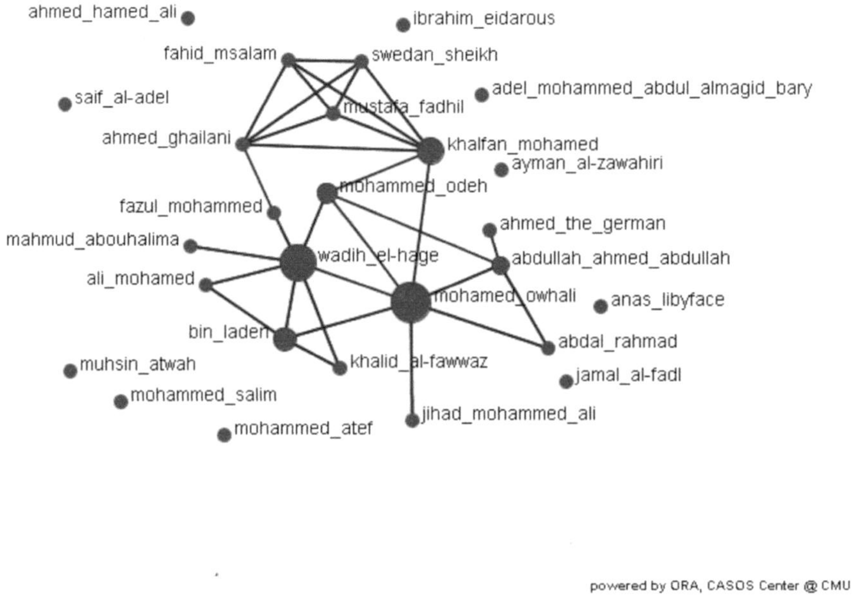


**FIGURE S-1** Graphical representation of a social network. SOURCE: Kathleen Carley, Carnegie Mellon University, Institute for Software Research International (2009).

communities. It is also dependent on the development of long-term intervention programs designed to change or improve conditions and behaviors in the community, making them resistant to stress and changes over time.

Discussion among workshop participants brought to light that many of the same capacities and characteristics that allow a community to continue functioning during a disaster (e.g., being well informed, well networked, and possessing the ability to respond to situations with creativity and flexibility) are those that allow a community to thrive during normal times. Many workshop participants stated that by increasing the capacity for effective communication through social networks, a community may be created that is resilient to a broad range of stressors. Investing in the building of community resilience is highly likely to yield rapid returns through the creation of stronger and healthier communities. According to many at the workshop, the application of SNA could advance resilience science and benefit community planning.

Emergency management practitioners who attended the workshop noted the need to establish measures of the effectiveness of disaster mitigation or response activities before establishing priorities and allocating resources. A "measuring stick" for social, economic, and relational capacities is of the utmost importance. However, because connections among organizations are not fully understood, the status of the connections cannot be measured, nor can they be measured for change. Workshop participants stated that baseline data could provide measuring sticks for changes in networks, the characteristics that foster community resilience, and the magnitude of realized or potential stresses caused by a variety of stressors. Quantifying which adaptive capacities are essential to community disaster resilience is necessary, according to workshop participants, in order to measure the effectiveness of activities to improve community disaster resilience.

Workshop participants expert in the field considered SNA theory and applications to be quite advanced, but participants stated that SNA is not being applied in ways that assist local communities and practitioners. From the scientific perspective, more and better data about networks are required for the development of the tools needed to advance the science and practice of SNA. Additionally, the means to test the validity of social science models resulting from SNA have yet to be developed. From the practitioner's perspective, explanations of SNA and its tools need to be made more meaningful to gain acceptance in everyday practice. Innovations and a proliferation of networking technologies (e.g., wireless technologies and networking software) are easily accessible. Awareness of both the positive and negative issues associated with the use of networking technologies to support social networking, however, would benefit the emergency management community.

## RESEARCH THEMES

Several research areas were identified by workshop participants as prerequisite to advancing the use of SNA for building community disaster resilience. Disaster management decision making depends on numerous factors including the phase of the disaster, available resources, and the level of authority at which decisions are made. SNA could improve situational awareness by emergency management practitioners by allowing them to understand and measure the status of networks within their communities. Using what is learned from SNA, necessary interventions and the conditions and network associations required for their success can be identified. The best means of communicating and implementing interventions can also be developed.

Numerous useful research topics were discussed by workshop participants and are described in the main body of this document. Recurring research themes discussed during the workshop are synthesized in the following sections. Workshop participants stated that addressing these themes could stimulate the use of SNA to build community disaster resilience. Barriers to conducting and applying the research are also discussed.

### Areas of Research

#### Baseline Data

Many workshop participants indicated that a certain level of baseline information regarding networks is necessary to determine the resilience of a community to extreme events. Baseline data describe the starting conditions by which change can be measured and include all manner of data regarding networks and their members. These data are fed into SNA to produce baseline models. Little, for instance, is presently known about who populates the formal, governmental networks responsible for a region's disaster management or how they may integrate with other social networks that reside in civil society for emergency management purposes. Without this baseline level of knowledge, it is difficult to evaluate the evolution of the composition of social networks and how these changes relate to resilience levels. However, collection and management of baseline data, according to participants, is not currently feasible given research funding practices. Workshop participants repeatedly stressed that the accuracy of network analysis, monitoring, and intervention design cannot be certain without baseline data.

### Validation Techniques

Mechanisms to validate new data, network models, and decisions made using SNA and related tools would also benefit practitioners and scientists. Practitioners described the need for mechanisms that can vet for accuracy the data traveling through a network, and indicate if the data require action or response. New networking technologies allow large amounts of data to travel quickly through networks. Practitioners need a means to sort which data are good, bad, redundant, and actionable.

### Understanding Network Dynamics

Networks are likely to change quickly during a disaster as infrastructure fails or is restructured, people relocate, or the availability of resources change. Building resiliency into social networks requires an understanding of how networks evolve during normal times, and during times of stress. Understanding how networks change when stressed, and how to promote positive changes that allow the networks to function during a disaster, is important. Some workshop participants suggested that new methods for studying network dynamics are needed. It is essential that network models be constantly updated. SNA tools would be more useful to practitioners if they allowed quick visualization of the changing nature and uncertainties in linkages within and between networks. This would allow more effective diffusion of information during all stages of the disaster cycle.

### Better Data Gathering Techniques

New and more refined data gathering techniques could result in better social network models. For example, workshop participants repeatedly stressed how SNA could be more effective if the means of obtaining proprietary and personal data for SNA while preserving the privacy of individuals and institutions were developed. Such data as who within and between private sector organizations communicates with whom and what kinds of people receive certain medical treatment under certain circumstances provide real insights into the nature of networks and their members. Workshop participants stressed the importance of maintaining privacy. Legal and ethical barriers are an issue.

### Government and Community Interaction

Workshop participants discussed that greater understanding of the ways individuals, organizations, locales, and agencies are connected to social networks and how these components are used would likely result in more effective use of networks to build community resilience. An understanding of how connectivity to networks may change under stress would also be valuable. Additionally, the skill sets and attributes of network members need to be understood in order to identify members that may emerge as trusted opinion leaders within their communities. These are individuals who could be enlisted to effectively disseminate information to their communities. Research on emergent behaviors—behaviors that arise as a consequence of a disaster—and the promotion of pro-response emergent behaviors among private individuals and individuals within organizations could also lead to better planning and the promotion of resilience.

### Exploring SNA in other Contexts

Study of how SNA is applied in areas such as network centric warfare, counter terrorism, and public health would aid in the application of SNA for improving community disaster resilience. The vocabulary of network-centric warfare is different from that used by social scientists, but the goals are similar: to understand and improve how information is sought and exchanged; and to develop action instruments that enable decision making. According to workshop participants, practitioners who collect, analyze, understand, model, and incorporate network data into their decision-making processes may be better poised to help their communities become more resilient. Building resilience is not only about preparation for disasters. Studying how networks deal with broader social issues would also be useful. Research on how communities deal with issues such as ethnic oppression may yield a rich and pertinent literature on community resilience from which to draw.

## New Research Paradigms

Barriers to SNA research and use of SNA tools by practitioners for building community disaster resilience were often discussed during the workshop. Although addressing these barriers is not directly part of the charge given the workshop planning committee, many participants noted that these barriers could affect the effectiveness of a future research agenda and the adoption of SNA tools in practice. Suggested strategies to overcome these barriers are summarized below.

Participants pointed out that current strategies for funding research and moving research results into practice are not adequate to address the large-scale and complex social science issues. New funding frameworks that accommodate larger and longer-term studies would benefit both the research and practice communities. For example, better baseline data from which progress can be benchmarked would probably result. Incentives to encourage rapid-response investigations immediately following an event, and multidisciplinary research in general, could lead to more immediately useful results for practitioners including information on topics such as intervention methodologies that have proven successful. Collaborative research conducted with practitioners, and between public and private entities, could make the adoption of SNA techniques among practitioners more likely. Removing barriers of access to infrastructure and data may also result. Workshop participants noted that the most relevant research, tools, and data for decision making would be those identified jointly by researchers and practitioners, with input received from the private sector.

Some practitioners and researchers at the workshop expressed concern that current homeland security priorities tend to encourage a focus on antiterrorism activities within the emergency management community. Some suggested that sources of community stress need to be adequately assessed to confirm whether a focus on antiterrorism is locally warranted. A better understanding of community stressors could allow for more informed allocation of resources.

Several workshop participants stated that researchers needed incentives to collaborate with practitioners. Placing more value within the university and research cultures in moving research into practice might foster such incentives. For example, the medical community has begun to support translational research (e.g., research on how to enhance the adoption of research products into practice) and translational activities (e.g., the training in the use of research results). These incentives have also encouraged universities to consider translational work in their decisions to award tenure to faculty. Adoption of similar policies in other research communities could encourage younger researchers (those most likely to be familiar with social networking technologies) to engage in translational work.

Workshop participants discussed the idea of developing regional collaboratives among local universities, agencies, and businesses. For example, local, state, and federal resources could be used to establish collaboratives to encourage thorough baseline expertise on regional social networks and adaptive capacities. The regions could be consistent with the 10 regions into which Federal Emergency Management Agency divides the United States. Each collaborative could be a repository for regional baseline data and serve as a resource for federal and local response agencies during crises. Longitudinal and rapid response investigations could tap those resources and be conducted within the collaborative framework.

1

# *Introduction*

A community's ability to respond to and recover from natural or human-caused disasters is in part dependent on the strength and effectiveness of its social networks. Effective interventions—activities designed to change or improve conditions in the community—during all phases of a disaster can be facilitated by community leaders taking advantage of existing social networks to send and receive information. Conversely, a community may be at risk if the relationships across the economic, cultural, social, and political sectors of the civic infrastructure are not understood. The ability to visualize interactions within and between community networks (for example, who communicates or works with whom) promotes situational awareness, rigorous coordinated planning, and the optimal allocation of resources necessary for disaster preparedness, community resilience, and response.

In sociological terms, community resilience is the ability of a community or social unit to withstand external shocks to its infrastructure (Norris et al., 2008). Community resilience emerges from the ability to adapt to stress and return to healthy functioning. The speed with which a community can mobilize and use resources during and following a disaster event is strongly dependent on its various capacities to adapt to change and is related to the strength of its social networks.

Social network analysis (SNA) is the study of complex human systems through the mapping and characterizing of relationships between people, groups, or organizations. Because SNA can reveal the characteristics, composition, and structure of existing networks, SNA may prove an important tool for understanding how the public and private sectors work together to respond to a disaster. SNA has been used to inform policy making in areas such as terrorism prevention, and could be useful for building community disaster resilience.

SNA could also be used to design or build networks for the purpose of building community resilience. The Human Factors Division within the Science and Technology Directorate of the Department of Homeland Security (DHS) applies social and behavioral sciences to support the preparedness, response, and recovery of communities affected by catastrophic events. Its goal is to advance homeland security technologies and planning by integrating human factors. The DHS contracted with the National Research Council

(NRC) to hold a two-day workshop to examine the current state of the art in SNA and its applicability to the identification, construction, and strengthening of networks within U.S. communities for the purpose of building community disaster resilience.

To answer its charge, the NRC formed an ad hoc workshop planning committee under the auspices of the Geographical Sciences Committee of the Board on Earth Sciences and Resources. The committee was tasked with organizing and conducting the workshop. The committee's statement of task is provided in Box 1-1. It includes the identification of elements of a future research agenda to support the design, development, and implementation of social networks for the specific purpose of strengthening the resilience of communities against natural and man-made hazards and terrorist events. The workshop took place February 11-12, 2009, and featured presentations and discussions on social networks, social networking tools, SNA theory and tools, and community resilience.

As described by Michael Dunaway of DHS, the ultimate result of a research agenda influenced by this workshop summary could be the creation of accessible tools that would enable county-level emergency management directors and other community leaders to define and visualize networks within their communities. With the ability to identify relationships within and among networks, social structures and adaptive capacities can be built and reinforced to make communities more disaster resilient. Workshop participants discussed whether such tools were possible and the type of research that could enable their development.

**BOX 1-1**
**Statement of Task**

An ad hoc committee will organize a two-day public workshop to examine the current state of the art in Social Network Analysis (SNA) and its applicability to the identification, construction, and strengthening of networks within U.S. communities for the purpose of building community resilience.

The workshop will explore the topic through invited presentations and facilitated discussions among invited participants, including the following issues:

- Current work in SNA that has focused on defining the characteristics, composition, and structure of existing networks (e.g., terrorist cells; infectious disease transmission; narcotics trafficking);
- Theories, principles, or hypotheses within the science of SNA that could be applied to the construction of designed networks to develop or enhance the strength of relationships within geographic or functional communities;
- Current research that has focused on the use of SNA for the development of designed networks;
- Gaps in current knowledge within the field of SNA that would inhibit the ability to apply SNA theories or principles to the construction of networks;
- Research areas that could fill gaps in this knowledge; and
- Elements of a research agenda that could be pursued to support the design, development, and implementation of social networks for the specific purpose of strengthening the resilience of communities against natural and man-made hazards and terrorist events.

## WORKSHOP PLANNING

### The Planning Committee

The workshop planning committee consisted of six members with expertise in the areas of SNA, spatial social science, hazards, resilience science, and community and disaster management. Appendix A provides biographies of the planning committee members. The committee held five teleconferences to discuss the statement of task, identify workshop participants, and develop an agenda. References shared among the committee members became the basis for the bibliography included in the workshop briefing materials and as Appendix B of this document.

### Structure of the Workshop

The workshop planning committee selected two major themes around which to organize the workshop: the use of SNA for preparedness and intervention, and the use of SNA in improvisational disaster response. The committee invited researchers and emergency management practitioners—those with their "boots on the ground" during an emergency—to participate. Researchers and community leaders from different geographical regions of the country, and with varying disaster experiences, were invited so that a broad range of issues and perspectives could be considered. A list of participants is presented as Appendix C of this document. Participants included individuals familiar with SNA for other purposes, such as identification of terrorist cells and for the development of programs to thwart the spread of infectious disease. Individuals that work with populations that could be disenfranchised during an emergency, such as the non-English speaking poor, were included among participants. The workshop agenda appears in Box 1-2.

The planning committee devoted the first morning of the workshop to defining topics to be discussed, including community resilience, social networking, and the states of the science and practice of SNA. Case studies in the use of social networks and SNA were provided. Summaries of presentations and discussions are found in Chapter 2. As indicated in Box 1-2, the introductory session was followed by concurrent breakout sessions on the major workshop themes, moderated by a member of the planning committee. The first set of concurrent sessions addressed how SNA could be used to enhance communication, and how SNA could be used for planning interventions in preparation for a disaster. The second set of concurrent breakout sessions addressed how SNA might enhance communications when coordinating the improvisational response of networks of organizations; and how SNA could enhance communication within local communities and among individuals. Breakout sessions concluded with a reconvening of workshop participants to summarize discussions. Appendix D includes descriptions of breakout session topics as well as questions developed by the workshop planning committee to guide discussion. Workshop participants were given these descriptions and questions before the workshop.

A concluding plenary session focused on key gaps in knowledge regarding the application of SNA to foster community disaster resilience and the research needed to fill them. Specific research themes were considered.

## WORKSHOP SUMMARY ORGANIZATION

This document summarizes the major points and ideas presented at the workshop as documented by a rapporteur. The summary reflects the specific topics emphasized by workshop presentations and discussions and may not be a comprehensive summary of all relevant topics and issues. Any documented observations contained in this summary are those of individual participants or groups of participants and do not necessarily represent the consensus of the workshop participants or planning committee, nor does the summary contain any consensus conclusions or recommendations.

This workshop summary is organized into four chapters. This chapter introduces the reader to the purpose and organization of the workshop. Chapter 2 summarizes the introductory presentations and discussions and explores the current states of the science and practice as presented by workshop speakers. Definitions of key terms used by workshop participants and in this summary are also provided. Chapter 3 summarizes the discussions of the utility of SNA in identifying networks and improving community resiliency before and during a disaster and in the response and recovery phases of a disaster. Chapter 4 synthesizes the ideas of workshop participants on how to move from the theoretical realm to the practical application of SNA for improving community resilience. Gaps in knowledge and potential research that could fill those gaps as identified by participants are summarized, as are barriers to SNA research and application.

**BOX 1-2**
**Applications of Social Network Analysis for Building Community Disaster Resilience: A Workshop**

**February 11-12, 2009**
**AGENDA**

***Wednesday, February 11, 2009***

8:30 Welcome and Introductory Remarks
*Susan Cutter*, Ph.D., Chair, Committee on Applications of Social Network Analysis for Building Community Disaster Resilience
University of South Carolina

**SESSION 1**
**INTRODUCTION: TOPIC OVERVIEW AND DEFINITIONS**
**(PLENARY)**

8:45 **Current State of the Art in Social Network Analysis**
*Kathleen Carley*, Ph.D., Carnegie Mellon University Institute for Software Research International

9:05 Discussion

9:20 **Fostering Community Resiliency: Theory and Practice**
*Fran H. Norris*, Ph.D., Dartmouth Medical School National Center for Disaster Mental Health Research

9:40 Discussion

10:10 **Reaching Vulnerable Populations through Social Networks**
*Carl Latkin*, Ph.D., Johns Hopkins Bloomberg School of Public Health

10:30 Discussion

10:45 **Using Social Networks to Enhance Communications**
*Michael Byrne*, ICF International

11:05 Discussion

11:20 **Synthesis and Discussion of Goals for Breakout Sessions**
*Susan Cutter*, Chair

**SESSION 2**
**USING SNA FOR PREPAREDNESS AND INTERVENTION**
**(CONCURRENT SESSIONS)**

1:00 p.m. **Session 2a: Communication**
**Moderator:** *William A. V. Clark*, Ph.D., University of California, Los Angeles
**Rapporteur:** *Kathleen Carley*, Ph.D., Carnegie Mellon University

**Session 2b: Planned Interventions**
**Moderator:** *Randolph H. Rowel*, Ph.D., Morgan State University
**Rapporteur:** *Monica Schoch-Spana*, Ph.D., University of Pittsburgh Medical Center

**Plenary Session**

3:45 Reports from Break-out Sessions & Wrap Up

4:30 Adjourn

***Thursday, February 12, 2009***

**SESSION 3**
**IMPROVISATIONAL DISASTER RESPONSE**
**(CONCURRENT SESSIONS)**

8:30 **Session 3a: Networks of Organizational Connections**
**Moderator:** *Monica Schoch-Spana*, Ph.D., University of Pittsburgh Medical Center
**Rapporteur:** *William A. V. Clark*, Ph.D., University of California, Los Angeles

**Session 3b: Networks within Local Communities and between Individuals**
**Moderator:** *Eric Holdeman*, ICF International
**Rapporteur:** *Randolph H. Rowel*, Ph.D., Morgan State University

**Plenary Session**

11:15 Reports from Break-out Sessions & Wrap Up

**SESSION 4**
**RESEARCH NEEDS AND IMPLEMENTATION GAPS**
**(PLENARY)**

1:00 p.m. What we don't know and need to know about SNA and resilience

- Identification of knowledge gaps and priority research areas
- Identification of specific research themes that enhance implementation of social networks as a means for increasing community resilience against disasters.

2:30 Next Steps: Moving the research agenda forward

3:45 Closing Remarks
*Susan Cutter*, Chair

4:30 Adjourn

2

# The Science and Practice

This chapter summarizes introductory presentations and discussions of the first session of the workshop. The purpose of the session was to introduce participants to the workshop charge (see Box 1-1) and the terminology to be used throughout discussions. Speakers were invited to provide basic information on the sciences and practices of social network analysis (SNA), fostering community resilience, reaching vulnerable populations through social networks, and the use of social networking tools to improve communication.

## WORKSHOP VOCABULARY

During the first day of workshop discussions, the workshop planning committee obscrvcd inconsistent use of key terms related to social networks and SNA by participants. To avoid confusion, the committee provided definitions for these terms (see Box 2-1). A social network is a group of people and organizations that form a web of relationships. Social networks were being confused with the tools used to facilitate them (such as Facebook[1]) or to analyze them. Social network analysis is the process of analyzing the key actors and connections within a social network. SNA can reveal redundancies and vulnerabilities within a network, and can be used to study the changes in all these variables. A product of SNA may be a graphical representation of a network that shows the interconnectedness of network members. An example is provided as Figure 2-1.

Issues were also encountered with the use of the term "resiliency." As described in Chapter 1, resiliency is the ability of a social unit to withstand external shocks to its infrastructure (Norris presentation to workshop participants).

[1]Facebook is a free-access, privately owned social networking website. See www.facebook.com (accessed March 24, 2009).

**BOX 2-1**
**Definitions of Key Social Network Terms**

The workshop planning committee developed the following definitions of key terms used in the study of social networks and social network analysis.

**Social network.** The interactions between people and organizations, including who knows, works with, or communicates with whom, that can be mapped. The data and information found, for example, in Facebook[a] and the Enron Email Corpus[b] are examples of social networks.

**Social network tools.** A set of computational techniques that enable individuals and groups to engage in social networking by monitoring and interacting within the networks with which they are connected. Facebook, MySpace,[c] and Twitter[d] are examples of social networking tools.

**Social networking.** The process of creating, maintaining, or altering one's network to one's advantage by using the network to gain resources or influence, or to mobilize activity.

**Social network analysis.** The process of analyzing a social network and identifying key actors, groups, vulnerabilities, and redundancies as well as the changes in these variables.

**Social network analysis tools.** The set of tools, technologies, metrics, models, and visualization techniques used for social network analysis. These may include data extraction tools, link analysis, statistical techniques, and graph theory techniques using programs such as AutoMap,[e] ORA,[f] UCINET,[g] and Pajek.[h]

**Social network theory.** The set of theories for forecasting, reasoning about, and understanding how social networks form, are maintained, and evolve, and the role of variables such as social networking tools, media, and stress in affecting the emergence, utilization, management, and change in social networks.

[a]See www.facebook.com (accessed March 2, 2009).
[b]See, for example, ebiquity.umbc.edu/blogger/2006/02/05/search-the-enron-email-corpus-online/ (accessed March 2, 2009).
[c]See www.myspace.com (accessed April 5, 2009).
[d]See twitter.com/ (accessed March 2, 2009).
[e]See www.casos.cs.cmu.edu/projects/automap/ (accessed May 10, 2009).
[f]See www.casos.cs.cmu.edu/projects/ora/ (accessed May 10, 2009).
[g]See www.analytictech.com/downloaduc6.htm (accessed May 10, 2009).
[h]See vlado.fmf.uni-lj.si/pub/networks/pajek/ (accessed May 10, 2009)

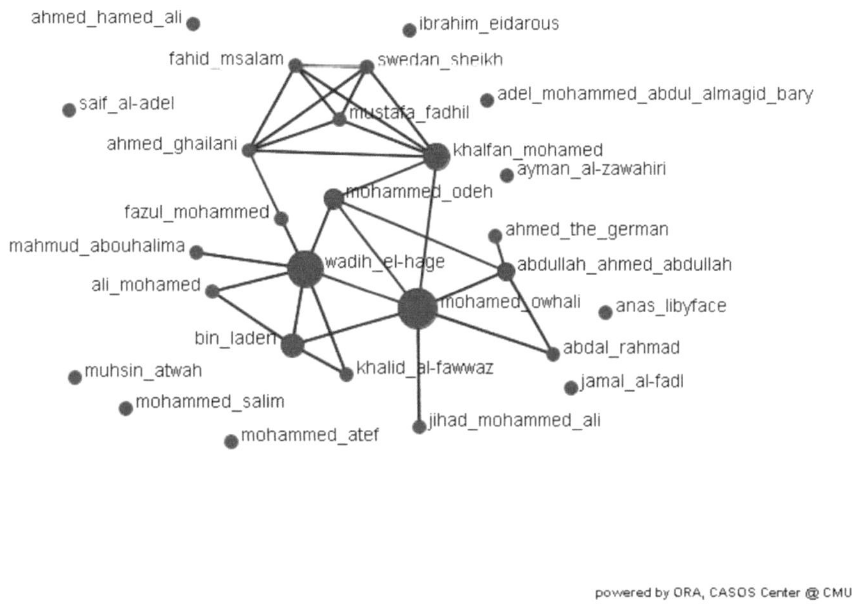


**FIGURE 2-1** Graphical representation of a social network. SOURCE: Kathleen Carley, Carnegie Mellon University, Institute for Software Research International (2009).

## THE STATE OF THE ART IN SOCIAL NETWORK ANALYSIS

Kathleen M. Carley of the Center for Computational Analysis of Social and Organization Systems (CASOS) of the Carnegie Mellon Institute was invited to give a presentation on the state of the art in SNA. She provided an overview of the main tools of SNA that focus on defining the characteristics, composition, and structure of existing social networks. Her presentation and subsequent discussion are summarized here. Other topics and issues may be relevant to SNA, such as the study of ego networks, but were not discussed at the workshop and not included in this summary. Unless otherwise noted, the ideas expressed in the following sections are attributable to Dr. Carley.

People, units of action, partners, departments, resources, ideas, skills, events, and countries can be graphically represented as nodes in a network (for example, the dots in Figure 2-1). The links—or ties—between the nodes are the interrelations and may represent physical ties such as roads or rivers, or less tangible ties such as alliances, associations, authority lines, transfer of resources, precedence, or who likes or respects whom. SNA can be used to identify and understand the relationships and strengths of the ties within a network, and understand how these ties are vulnerable under certain circumstances. SNA can also be used to conduct sentiment analyses to understand the flow of ideas or feelings.

Classic social network and link analysis modeling and basic pattern detection capabilities are readily available but not commonly applied in policy making. Nonetheless research in SNA is rapidly expanding and cutting-edge technologies yield exciting results with sociopolitical ramifications. SNA is mostly unused at the local level with the exception of epidemiological studies, such as the tracking of disease sources, and in counterterrorism and counternarcotic investigations. Most of what is known about social networks, however, is not integral to disaster management practice.

SNA for any purpose is often thwarted by the discrepancy between the amount of data needed to yield meaningful results and the amount of data available. The lack of technological capacity and the lack of social science skills needed to correctly apply SNA-derived models are also factors. Researchers addressing the data question do not expect to close this gap within the next decade.

## Cutting-Edge Approaches: Dynamic Network Analysis

Traditional SNA focuses on nodes within a network and considers the attributes that make an individual node stand out. The state of the art in the field of SNA is beyond determining who communicates with whom. Newer approaches to SNA consider networks as a whole, and powerful techniques exist that allow the analyses of the what, where, how, why, and when of situations. These techniques enable a user to identify the need for interventions, plan for them, and provide input for policy management. Most agencies, including disaster management agencies, currently collect "trail data," such as who entered a health department on what day for what information, or who crossed the border at what time.

State-of-the-art data collection includes a dynamic network analysis suite of three types of tools to track and analyze trail and other network data. The suite includes (1) data mining tools, such as AutoMap, that collect network data from open sources such as newspapers to identify network components; (2) statistical analysis-type packages, such as Organizational Risk Analyzer (ORA), that take into account social and dynamic network metrics, conduct broader data mining and link analyses, and apply machine learning techniques for clustering; and (3) simulation tools, such as DyNet, Construct, and BioWar, that allow scenario analysis for the consideration of various options. Box 2-2 provides descriptions of some of these tools.

Although dynamic social network analysis is used in some applications, many barriers exist to their widespread use for analyzing complex networks. It is difficult to extract text and links from the wide range of required data sources. Many applicable analytical techniques for city-scale networks require extensive computational resources. Additionally, many simulation models are built for a single purpose and cannot be reused, quickly making them obsolete. Finally, interpreting and moving this level of information into the policy realm and scale is not yet a functional reality.

**BOX 2-2**
**Dynamic Network Analysis Tools**

**AutoMap** is a product of CASOS at the Carnegie Mellon University and is a text mining tool that enables the extraction of network data from texts. The tool can extract content analytic data (words and frequencies), semantic networks, and metanetworks. The main functions of AutoMap are to extract, analyze, and compare mental models of individuals and groups, and to reveal the structure of social and organizational systems from texts.
SOURCE: www.casos.cs.cmu.edu/projects/automap/ (accessed March 21, 2009).

**BioWar** is a CASOS package that enables community leaders to prepare for biological attacks using computational models of social networks, communication media, disease models, demographically accurate agent modes, wind dispersion models, and a diagnostic error model combined into a single model of the impact of an attack on a city.
SOURCE: www.casos.cs.cmu.edu/projects/biowar/ (accessed March 21, 2009).

**Construct**, developed by CASOS, is a multiagent model of group interactions where agents communicate, learn, and make decisions in a continuous cycle. The program takes into account how agents learn through interaction and change their perception of the environment.
SOURCE: www.casos.cs.cmu.edu/projects/construct/info.html (accessed March 21, 2009).

**DyNet** is a reasoning support tool developed by CASOS intended to simulate reasoning about dynamic networked organizations under varying levels of uncertainty using computer science, social network, and organization theory.
SOURCE: www.casos.cs.cmu.edu/projects/DyNet/dynet_info.html (accessed March 21, 2009).

**i2 Analyst's Notebook** is a commercial visual investigative analysis tool that allows investigators to organize large volumes of disparate data and conduct link and timeline analyses.
SOURCE: www.i2inc.com/products/analysts_notebook/ (accessed May 12, 2009).

**Organizational Risk Analyzer (ORA)** is a risk assessment tool developed by CASOS that examines network information and identifies individuals or groups that are potential risks to a network given social, knowledge, and task network information.
SOURCE: www.casos.cs.cmu.edu/projects/ora/ (accessed May 12, 2009).

**Palantir** is a commercially available information analysis platform for integrating, visualizing, and analyzing structured, unstructured, relational, temporal, and geospatial data for security, intelligence, defense, and financial applications.
SOURCE: www.palantirtech.com/ (accessed May 12, 2009).

**R** is a computer language and environment for statistical computing and graphics developed by Bell Laboratories. SOURCE: www.r-project.org/ (accessed May 12, 2009).

**Starlight Information Visualization System** is a visualization-oriented user interface for temporal and spatial information analysis and network modeling developed by the Pacific Northwest National Laboratory. SOURCE: starlight.pnl.gov/ (accessed May 12, 2009).

**UCINET** is a commercially available comprehensive package for the analysis of social network data using a variety of network and statistical analysis methods.
SOURCE: www.analytictech.com/ucinet6/ucinet.htm (accessed May 12, 2009).

## Functional Applications in Dynamic Social Network Analysis

Available tools make it possible to conduct network analysis with open-source, raw-text data input, such as information from newspapers, and then conduct scenario analysis (e.g., what would happen given a certain set of circumstances), and finally conduct analysis that can identify emergent leaders. Analysis is possible, for example, that can connect all of the potential emergency responses in a community to specific emergency responders in order to see where vulnerabilities in a response network exist. Location analysis can be done to see how things are done differently in different areas. Geospatial network analysis is possible, as is information gain and loss tracking. It is also possible to detect changes in organizations and behavior over time within a network using available analytical techniques. If the resources are available, the mapping of belief structures and trends over time is possible and may allow policy makers to identify where people hold certain beliefs, where beliefs are likely to change, who the critical actors are that enable change, and to predict who will be central to the network in the future. Belief forecasting analysis can be conducted for given types of network structures and sociodemographics. The results of the analysis provide policy makers and leaders with powerful information to help them determine how best to communicate with and enlist the assistance of their communities.

The resulting data and models from dynamic social network analyses can allow managers to identify critical network features, identify opportunities for intervention analysis or action, and conduct limited types of event forecasting.

## Issues and Knowledge Gaps in the Application of SNA

Traditional SNA technologies that are able to reveal weaknesses in response networks, identify vulnerable populations, target opinion leaders in communities, or conduct text mining to support hot-topic analyses are not regularly utilized in policy-making settings. This is also true in the disaster management community where, in general, networks do not exist that link emergency responders with one another or with networks elsewhere in the community. Some workshop participants expressed the view that SNA could be applied in analysis of the emergency management community and emergency response plans that are in place at the national, state, and local levels.

### Proper Use of Tools

Framework modeling and network statistical analysis tools are readily available to community and disaster managers, but those using them often are not familiar with community social science models. Under such circumstances, statistical analyses may be overapplied, good interpretation of network situations may be missing, and resulting models may be in error. Even scientifically sound network models may be used incorrectly, or metrics for change may be misinterpreted. According to some workshop participants, increased communication between social scientists in the research community and modelers within emergency management communities would be beneficial. A barrier to collaboration, however, is that researchers and practitioners do not use the same analysis tools. Different tools are used, in part because of the cost and accreditation of software, and because of the scalability and visualization capabilities of various software packages.

## Need for Translational Research

To realize the benefits of SNA, it is essential that information move from the research realm into practice, also known as translation of research. Research conducted on the best means to translate information is defined as translation research. Translation research can be helpful in developing an educational process that demonstrates how the adoption of new ideas and tools will yield actionable results for practitioners. Methodologies, language, and examples that would be most meaningful to a target audience can be identified. Louise Comfort of the Graduate School of Public and International Affairs, University of Pittsburgh, described experiences promoting the use of networking technologies in different municipalities. She found that new technologies could be quickly accepted if individuals were willing to think beyond their traditional routines (Comfort and Wukich, 2009). She described the interaction and process of learning that occurs on at least three levels to create a learning network:

(1) Individuals learn the technologies;
(2) Efficient and easier communication occurs among individuals as a result; and
(3) Collaboration and validation of information occurs.

Generally, younger personnel with access to better equipment were more willing to accept new technologies. Acceptance of technologies into practical use has occurred within the drug-traffic enforcement and healthcare communities. Acceptance more often occurs when champions of the technologies are identified within the communities to collaborate in the development of educational materials. Similar relationships would be useful to the disaster management community.

Translational activities validate basic findings for practitioners in language that can be understood, and can result in the decrease in the time to move a concept from the research realm into practice. Dr. Carley indicated that the engineering field engages in translational research that results in a relatively small lag of six years between the inception of an idea and its practical application. She stated that the translation of complex SNA techniques is estimated to occur only after approximately 24 years because the SNA community is not actively engaged in translational research and activities. Without translation research, widespread benefits resulting from the application of SNA will be negligible. In spite of this, no agency is charged with funding such research.

## Cost

There are different but related costs to be factored in by a community when considering SNA technologies for building community disaster resilience: the cost of the necessary analytical tools, the cost of creating the network of individuals to conduct the analyses, and the cost of creating the community networks necessary to develop community resilience. It was discussed during the workshop that a complex network analysis for a system at the city level could require between half a million and several million dollars. The cost can vary significantly depending on the data already available and the level and condition of available hardware. The cost of SNA tools may be controlled by taking advantage of free and currently available state-of-the-art tools. Agencies typically use commercially available software at a cost of thousands of dollars.

### Validation of Models

According to Dr. Carley, confidence in models developed using SNA tools is necessary before policy makers will make model-based decisions. However, global datasets essential to validating models do not exist. This is particularly true in the area of disaster preparedness for which large-scale baseline or control data for comparison to projected models are not available. Detailed data may be available regarding specific investigations, for example, arrest records for research documenting specific crimes following a disaster, or health care records documenting a specific disease outbreak. These data, however, are often incongruent, not comprehensive, and not global in scope. When combined they can often lead to baseline models that are inadequate. Legal barriers and unwillingness of agencies or jurisdictions to share data factor into the unavailability of data. Privacy and security issues are a primary reason for this unwillingness to share. A workshop participant indicated that this issue was recognized and discussed in a recent NRC study *Successful Response Starts with a Map* (2007).

The only standards available with which to validate complex social system models are engineering standards. These are not adequate for the task. New technologies for social model validation could result in reduced error and better models.

### Behavioral Factors

Not enough is currently understood about how trust in and reliance on information sources change as a result of stress. A better understanding of the nature of these changes in a technology environment could allow these concepts to be usefully incorporated into network models and decision making. Additionally, researchers may understand how data are collected off the Web and how individuals use their networks, but it might not be understood how the flow of information changes if the status of individuals' connectivity to the Web changes. Researchers do not know, for example, how Internet penetration in a network changes who the opinion leaders of a network are. This gap is acknowledged and being addressed by the Office of the Secretary of Defense.

## STUDYING, ASSESSING, AND CREATING RESILIENT COMMUNITIES

Fran H. Norris of the National Center for Disaster Mental Health Research of the Dartmouth Medical Center was invited to define community resiliency in a presentation to workshop participants. This section summarizes her presentation in combination with the discussion it generated. Unless otherwise noted, conclusions may be attributed to Dr. Norris.

## What is Resilience?

Resilience can be understood as a response to stress and can be considered as (1) a theory that guides the understanding of stress response dynamics; (2) a set of adaptive capacities that call attention to the resources that promote successful adaptation in the face of adversity; and (3) a strategy for disaster readiness against unpredictable and difficult to prepare for dangers. Response to stress can occur at the individual, institutional, or societal levels. The overarching concept is one that can guide research, policy, and the design and sequencing of interventions. Building community disaster resilience is more than a focus on disaster preparedness. It represents a paradigm shift that relies on building economically strong communities whose members can work together and use information to make and act on decisions.

The concept of resilience is familiar in many disciplines. In physics and mathematics, resilience refers to the speed with which a material or system returns to equilibrium after displacement. In ecology, resilience refers to the persistence of relationships within a system and the ability to absorb change. In psychological terms, resilience refers to the process of successful adaptation despite challenging or threatening circumstances. In sociology, resilience is the ability of social units, such as communities or cities, to withstand external shocks to their infrastructure. In all these definitions, resilience involves a process and reflects adaptability rather than stability. Resilience is not an immutable condition but rather a set of adaptive capacities to be continuously attended to and modified to prevent their loss. A community that functions well, consists of members that are behaviorally and mentally healthy, and offers a high quality of life is more likely to be able to adapt after a disturbance or adversity.

## Resilience as a Trajectory of Adaptation

Resilience is one of multiple possible stress responses to a disaster that also include resistance, recovery, and chronic dysfunction. Communities resistant to a specific event are barely affected by it. A resilient community may display transient dysfunction that is quickly resolved following an event. For some communities, dysfunction is more slowly resolved, but the community ultimately recovers. Chronic dysfunction represents the failure to adapt to the new circumstances. Factors controlling response trajectories are severity of exposure, the existence (or the perception of the existence) of social supports, and social class.

Different levels of distress can be observed within a single community following a disaster, as demonstrated in a case study of mudslide victims in Mexico in 1999. A high degree of property damage, bereavement, and complete displacement was observed following mudslides that destroyed a large portion of a community. Levels of distress, such as symptoms of post-traumatic stress disorder, were monitored among the population over time. Approximately one-third of the community was resistant to stress and was observed to have few symptoms. Another third of the community showed very high initial distress that improved at different rates. This group displayed different levels of resiliency but eventually returned to normal function. The remainder of the population showed moderate and high levels of distress that did not improve over a 24-month period. This group was considered chronically dysfunctional.

## Building Adaptive Capacities to Increase Community Resilience

Community resilience emerges from the ability to withstand stress without degradation. It is largely dependent on access to vital community resources. The rapidity with which resources can be accessed and used during and following disruption contributes to resiliency, as does redundancy of vital community elements in the case of failure of an individual element. Community resilience is more likely to result in the building and balancing of different qualities related to:

- **The social capital** available to the community through its networks, including organizational linkages within a community, the amount of social embeddedness, the attachment to place, the sense of community, citizen participation, and the real and perceived support in the face of adversity;

- **Community competence**—a measure of how ordinary people make decisions. Community competence is dependent on actions of the community, community problem-solving skills, flexibility, creativity, mutual trust in the effectiveness of people working together, and the belief that the community is empowered to control resources; and

- **Information and communication,** including competent communication skills and infrastructure, and trusted, responsible, and relevant sources of information.

Balanced levels of social capital, community competence, and ability to access and communicate information build resilience by enabling those qualities that help a community function as a community in the face of adversity. Economic development—including the level and diversity of economic resources, the equity of resource distribution, and the fairness of distribution of risk and vulnerability to hazards—is also important in building resilience.

Significant improvements in community resilience could result using the public health approach of encouraging small shifts in population response to disaster events. Figure 2-2a shows the distribution of the common response trajectories of victims of the Mexican landslide case study described earlier. Figure 2-2b shows what the responses in the same population could be if only 5 percent of the total population could be made more resistant, 5 percent could be made more resilient, and 5 percent of those who would not have recovered could be assisted into recovery. Such improvements can be achieved by intervening at multiple points, and addressing multiple adaptive capacities, before and during a disaster, and again later in time. Such improvements may include

- Developing economic resources, reducing of resource inequities, and giving attention to the areas of greatest social vulnerability;

- Meaningfully engaging local people in all steps of mitigation processes;
- Fostering organizational relationships to rapidly mobilize needed services;
- Boosting and protecting naturally occurring social supports;
- Planning for the unexpected by exercising flexibility; and
- Building trusted information sources that can continue to function in the face of unknowns.

The means to measure adaptive capacities, especially in the area of communication, have not been developed. SNA may provide a reasonable means of measuring the adaptive capacities associated with community resilience and in determining how best to intervene to achieve the desired improvements.

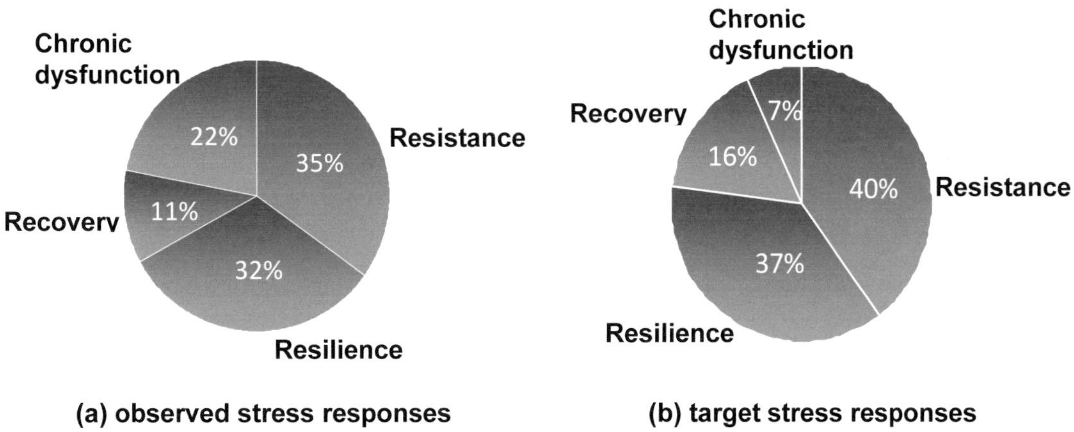


**FIGURE 2-2** Distribution of response over a 24-month period by a population impacted by Mexican mudslides in 1999. 2-2a represents observed responses. 2-2b represents the potential response given a positive shift of 5 percent in both the resistant and resilient groups, and in those able to recover in the longer term. Note the percentage of the population that remained chronically dysfunctional would be dramatically reduced with the application of this public health approach. SOURCE: Fran Norris, Dartmouth, workshop PowerPoint presentation.

## REACHING VULNERABLE POPULATIONS THROUGH SOCIAL NETWORKS: CASE STUDIES OF EFFORTS TO PREVENT THE SPREAD OF HIV

The workshop planning committee sought to explore how social networks and SNA could be used to reach vulnerable populations that may become disenfranchised from the larger community following a disaster. Because of the limited number of case studies on this topic, the committee looked to a case study from the public health community. Carl Latkin of the Johns Hopkins Bloomberg School of Public Health was asked to give a presentation on his efforts to recruit inner-city residents at high risk for HIV infection and transmission to promote positive behaviors within their communities. His presentation was related to how social and personal network approaches were used to both create networks among vulnerable populations and potentially influence behaviors within the networks. In the social network of impoverished inner-city populations, a high degree of network linkages exist among impoverished individuals and individuals with chronic physical illnesses, mental illnesses, and drug addiction. Consequently, social network analyses afford a viable approach to reaching these vulnerable populations.

A summary of Dr. Latkin's presentation and ensuing discussion are provided here. Conclusions presented are by Dr. Latkin unless otherwise noted.

## The Urban Social Service Network Setting

Insufficient service capacity exists to deal with the demands of daily medical emergencies in many major northeast urban cities. Emergency Medical Services (EMS) may be overwhelmed by nonemergency uses by substance abusers, the mentally ill, those in nursing homes, and the homeless. The design of social service networks discourages use of public resources by forcing people to wait in long lines, by treating them poorly, and encouraging distrust between service providers and impoverished populations. Building successful networks and services is dependent on building trust, an important adaptive capacity. Interacting with EMS or the fire department is a good way to see the conditions, plights, and social isolation before developing interventions.

## Understanding the Network

The Johns Hopkins Bloomberg School of Public Health used network approaches to reach highly impoverished individuals and change unhealthy behaviors. Inner-city residents at high risk for HIV infection and transmission were recruited in Philadelphia, Pennsylvania, and in Thailand to promote risk reducing behaviors in their communities (Latkin et al., 2009). Systematic study of and establishment of rapport with the communities were necessary to gain the trust vital to the success of the programs. The nature and stability of needle-sharing networks over time were studied. Networks consisted of up to 10 individuals who shared needles and sex. Some network stability existed, but a fair amount of fluidity and turnover of individuals were observed. Some linkages of these smaller individual networks to a larger social network were noted, but not all members of the community were linked.

## Introducing Interventions

Individuals with a desirable combination of skills and natural leadership abilities were enlisted from over 400 of the needle-sharing networks to educate their peers. They were 18 years or older, had weekly contact with active drug users, and were willing to be trained to conduct outreach education and bring network members into the program. The identified leaders were paid to receive training, but received only symbolic rewards for talking to their friends, modeling risk reduction behaviors, and socially influencing the critical behaviors that prevent HIV infection. Given the high level of stressors and traumas within these populations, redundancies were incorporated into the networks because of the likelihood a leader would become unavailable. Control groups were established to measure the effectiveness of information diffusion and the potential of behavioral changes.

Over a thousand community members participated in the intervention that consisted of six small-group peer-educator training sessions. After 24 months, participating network members were up to twice as likely to have reported not engaging in high-risk injection behaviors. These individuals were also more likely to have engaged in conversations regarding HIV risk behaviors following the training. The efficacy in reducing risk was not established in the studies.

## Issues Related to Designing Interventions

Recruitment methods to identify program participants may substantially influence interpretation of network structures and functions. People identified as central to networks may not be interested in participating in the interventions or may not be the best opinion leaders. Negative reactions may result if a message is received from a source lacking credibility. There may be role conflict for a member of the community representing the program. Individuals need to be trained in how to maintain credibility. Lack of resources within networks is a barrier to effectiveness, but changes in behavior can be associated with a small amount of resources.

Workshop participants discussed how interventions have to be carefully planned and translation efforts targeted to be robust and sustainable, especially given the lack of control over the message once it moves into the network. Information has to be disseminated in an appealing and memorable way to withstand mutations as it spreads.[2] Using rumors (described as the "grapevine" by workshop participants) to translate information can be an effective way to make messages more durable. Narratives that give experiences shared meaning and purpose are important.

Inherent in attempting to change behaviors of individuals in a network is the possibility of changing the network itself. In the case studies presented, individuals whose risk behaviors changed the most were also more likely to drop their ties with the network. This is something to consider when designing intervention networks. How the role of the trained leader is maintained or transitioned at the end of the intervention program so that the message is sustainable is another consideration.

## Issues Related to Disaster Preparedness

Concentrated in many impoverished inner-city neighborhoods is a phenomenally high level of drug and alcohol addiction, chronic disease requiring medication, and mental illness. Individuals afflicted by these ailments often have only the care of others similarly afflicted on which to depend. Community resources are unavailable or not trusted by the population during times of normal operation, but are even scarcer when disaster strikes.

---

[2]For example a more durable approach to informing drug users of the effectiveness of cold water on needles may be to compare the use of cold water on needles with using cold water on clothing to wash out blood.

The case studies presented here describe intervention efforts targeting small networks whose members may not be attached to larger community networks. Workshop participants discussed that the stability of these networks is fragile at best, and disintegration of the networks is likely following a disaster. Individuals within these communities are at severe risk of being further disenfranchised from the larger community, and may lack any knowledge of, access to, or trust in aid offered in response to a disaster.

Workshop participants expressed a need to consider how communication with all members of a community, including individuals within disenfranchised or potentially disenfranchised populations, should occur. The means to communicate and provide services to all community members during each phase of the disaster cycle is essential. Special study of the means to build disaster resilience among fragile communities, such as those described by Dr. Latkin, is desirable.

It is important to go into communities to understand the access, feasibility, and reliability of resources, and to understand how many people are reliant on the same resources. This is especially true in disaster management settings when many people may depend on the same resources, or the availability of the resources may change. Social network analyses afford a viable approach to reaching these vulnerable populations.

## USING SOCIAL NETWORKING TOOLS TO ENHANCE COMMUNICATION

In the last decade, numerous digital networking tools have been developed that are changing the way many in American culture communicate. The workshop planning committee invited Michael Byrne of ICF International Inc. to provide background on these tools and relate how emergency managers could use them during all phases of a disaster. This section summarizes his presentation, entitled "Impact of Technology on Collaborative Homeland Security: Web 2.0, 3.0, 4.0 and Beyond," and the discussion that followed. Many of the experiences relayed were anecdotal but are descriptive of the issues and solutions at hand.

A new vision for the Internet began to take shape during the Web 2.0 Conference in 2004.[3] Web 2.0 represents a culture shift, with the Internet being controlled by users from the bottom and providing an interactive environment that fosters innovation. Users become active participants rather than observers. The Internet now offers rich user experiences including Web video, interactive maps, timely content, and virtual worlds[4] which can be used not only for online entertainment but also for practical purposes such as conferencing and training. The evolution of the Internet will continue beyond Web 2.0 with the development of tools such as autonomous intelligent agents that are programmed to recognize user interests and filter and manipulate information the user sees. This is already being applied to some extent to target advertisements to Internet users.

As defined in Box 2-1, social networking tools enable individuals and groups to engage in social networking by monitoring and interacting within their networks. Text

---

[3]A conference held by O'Reilly Media Inc. and MediaLive International (www.web2con.com/web2con [accessed April 5, 2009]).

[4]Virtual worlds are computer-based simulated environments in which single or multiple users can communicate and manipulate events within the environment.

and multimedia information can be easily shared using relatively inexpensive and accessible technologies and distribution networks available for free on the Internet. According to Mr. Byrne, 3.75 billion people in the world have mobile communication devices. Sixty-two percent of all Americans have experience accessing wireless digital data and tools (Horrigan, 2008). Emergency managers who do not use these tools to reach their communities, build networks, and improve communication risk the possible detriment of their communities.

Twitter allows members to distribute text messages, called "tweets," of up to 140-character with their cell phones to geographically-, group-, or friend-based networks. According to the website TechCrunch,[5] information regarding the 2008 terrorist attacks in Mumbai was shared worldwide in quasi-real-time using Twitter—faster than news agencies such as CNN reported the events. In London, 62,000 cameras record much that occurs in public spaces. Communities in England are networked, and video information of interest can be shared in real-time. Individuals anywhere in the world can share visual information over networks using tools such as Flickr[6] and YouTube.[7]

The emergency management community has largely missed the networking revolution. Emergency management practitioners would benefit from a new communication paradigm and from studying how others are using social networking tools. For example, the Department of Health and Human Services has used the virtual world Second Life to run training drills;[8] law enforcement agencies have used gaming technologies for "shoot, don't shoot" drills; and America's Army[9] uses gaming technologies to teach basic first-aid skills. This type of training could prove to be an inexpensive alternative to traditional training approaches.

Workshop participants heard anecdotal evidence that evacuated New Orleans city staff used social networking tools to organize and get things done following Hurricane Katrina. This was possible in spite of individuals being spread geographically, and far from New Orleans. The electronic convergence of people into widespread information networks can enhance the concept of resiliency in a global sense.

## Digital Divides

Multiple digital divides exists among users of social networking technologies. Currently, social network tools are used most actively by people younger than 25 and older than 50 years (Li and Bernoff, 2008). Fifty-five percent of online teens have created profiles on social networking websites but far fewer online adults have done so (Madden et al., 2007). Other divides exists because of lack of financial or technical resources. Whatever the reason for the divides, penetration of networking technologies in society is far from complete and the incompleteness can be socially stratifying. The most vulnerable populations during and following Hurricane Katrina were those least likely to

---

[5]See www.techcrunch.com/2008/11/26/first-hand-accounts-of-terrorist-attacks-in-india-on-twitter (accessed April 3, 2009).
[6]Flickr is a Web-based photo management and sharing tool (see www.flickr.com [accessed April 3, 2009]).
[7]YouTube is a video sharing website where users can upload, share, and view video clips (see www.youtube.com [accessed April 4, 2009]).
[8]See secondlife.com (accessed April 5, 2009).
[9]See www.americasarmy.com (accessed April 5, 2009).

use or have access to social networking tools. It is still essential to conduct door-to-door searches following an emergency in spite of great advances in networking technologies.

A digital divide also exists between many government entities and the public. Many organizations prohibit access to Internet sites such as Facebook and MySpace.[10] The organizations have not kept up with networking technologies or have not used them advantageously to communicate with constituents. Government agencies will be unable to ignore networking tools because of the growing reliance on networking (versus traditional communication approaches) by the public. Many municipalities resist the adoption of networking tools because of valid or perceived security concerns. Agencies at all levels often opt to take control of security issues by creating their own networking tools—an expensive approach that could result in quickly outdated technologies. However, some government agencies are now exploring and even embracing the active use of social networking tools.

## Double-Edged Sword

Even though networking tools can be successfully used to spread information, Mr. Byrne also sees them as a double-edged sword. Information may not be well managed, systems may be overwhelmed by a large number of messages, it is difficult to prevent the spread of misinformation, and infrastructure can fail during catastrophic events. High-tech solutions need to be balanced with lower-tech solutions to ensure that redundancies and backups exist. Network tools are easily used by those with both honest and dishonest intent. Workshop participants learned how some terrorist organizations are known to use tools such as Twitter and Google Earth[11] to update their networks. The challenge for the emergency management practitioner is to synthesize and analyze the large volume of information available and determine whether the information is correct, actionable, or requires response.

Some believe that a large group of people sharing information can arrive at more accurate conclusions than a small group of experts discussing a given topic (Surowiecki, 2004). Many believe this is the strength of the Internet. Interactive connectivity implies constant feedback that makes a system self-correcting. As of 2007, there were over 60 million blogs[12] on the Internet (Wyld, 2007). Sites such as YouTube and Facebook create a value beyond what a top-down control model provides. However, bottom-up organizations largely shaped by its members can be disorganized, loosely controlled, unmethodical, and sometimes inconsistent, in part because they are in constant states of flux. Conflicting information may make it difficult to determine which information sources are accurate, and bad information may promote unwanted behaviors. Charles Mackay wrote a book in 1841 entitled *Extraordinary Popular Delusions and the Madness of Crowds* in which he discusses the perils of the spread of misinformation. The high

[10]See www.myspace.com (accessed July 16, 2009).

[11]Google Earth is a virtual on-line globe with free downloadable drivers that displays satellite images of the earth's surface at different resolutions. Users can add their own data and overlay their own images (see earth.google.com/index.html [accessed April 3, 2009]).

[12]A blog is a user-generated and regularly updated online journal. An example blog mentioned by Michael Byrne is *Disaster Zone: EmergencyManagement in the Blogosphere,* maintained by Eric Holdeman (see www.disaster-zone.com [accessed April 3, 2009]).

level of connectivity in society today creates the potential for major disasters or magnification of disasters through the unintentional or intentional misuse of networking tools.

Bad information can be long-lived on the Internet and can persist even on successful social institutions such as Wikipedia,[13] where entries are subject to constant review by members. Even considering content error, these institutions remain successful because within their bottom-up organizational structures, ways are available to manage data, look for and fix problems, and recognize and resolve attacks on the system.

For application in the disaster management community, it is essential that systems and networks are functioning before a disaster in order for them to be effective during and following a disaster. The physical infrastructure required to operate the networks may be resilient, but infrastructure failure is possible during a catastrophic event. In this situation it is essential to assess the needs of the impacted community and to communicate to people outside the affected area about the contributions they can make. It is also essential that the physical infrastructure be restored.

Workshop participants pointed out that emergency response plans generally call only for restoration of communication infrastructure among response agencies, and no authority exists among emergency managers to restore communication and networks used by the public. However, current networking technologies allow the quick localized re-emergence of networks wherever a transmitting device is functioning. Some management of network re-emergence would benefit emergency managers. Partnerships among public and private entities could provide solutions.

## Shifting Paradigms

Workshop participants discussed how emergency management organizations within the United States do not employ a command and control structure for disaster management. Instead, a kind of diplomatic structure is in place. This structure can be enhanced and made more effective using networking tools to build community networks and improve disaster resilience. Technology now allows emergency managers to shift from a focus on what needs to be done for the community, to what can be done with the community. There are an increasing number of examples of how emergency managers are using networking tools to change the mindset of community members—changing them from victims to survivors and eventually to community resources.

Inexpensive networking tools and devices used by a large proportion of the population. Emergency managers can engage an existing audience, and there is a great potential for positive outcomes and for lives saved. Positive outcomes will only be achieved, however, with an understanding of how a population will behave in response to new information, how a message may change once it is broadcast, and how durable messages are developed to have the desired effect. Many workshop participants expressed the importance of understanding the effective use of social networks and social networking tools during all phases of the disaster cycle while remaining cognizant that the infrastructure for these tools may fail. Understanding how to apply networking

[13]Wikipedia is a free, Web-based encyclopedia with entries created, collaborated on, and maintained by volunteer users (see en.wikipedia.org/wiki/Wikipedia:About [accessed April 3, 2009]).

innovations at different social and management levels to identify and close disconnects between those that need resources and those that have access to them would be beneficial. Networking tools can be intelligently used to engage meaningfully with the community to positively influence behaviors, influence mitigation and preparedness strategies, build more resilient systems, and improve response and interventions.

3

# *Social Network Analysis for Improved Disaster Preparedness and Intervention Planning*

Chapter 2 summarized presentations and discussions of the first session of the workshop regarding case studies and real-world examples of social network analysis (SNA) application. This chapter summarizes Sessions 2 and 3 of the workshop (see Box 1-2 for workshop agenda). Session 2 was devoted to exploring the use of social networking and SNA for disasters and public health emergencies; it was focused on the empirical identification of social support networks, the measuring of networks and understanding communication among them, and on assessing opportunities for interventions to improve community disaster resilience. Workshop participants broke into two groups that met concurrently. Session 2a addressed the role of communication and ways to improve the functional, structural, and interactional linkages between networks to foster preparedness. Session 2b examined the opportunities and constraints in the application of SNA for planning interventions that improve community disaster resilience.

Session 3 was devoted to exploring the use of social networking and SNA for improvisational disaster response. Opportunities to use SNA to facilitate emergency preparedness and response were examined for two different venues: networks among and between organizations; and networks within local communities. As in Session 2, workshop participants were broken into two groups that met concurrently. Session 3a addressed how SNA could be used to make interactions among networks of organizations more productive and improvisational responses more flexible. Session 3b focused on how individuals and communities could be empowered and engaged to foster collective behavior while preparing for, responding to, and recovering from disasters.

As described in Chapter 1, questions for guiding discussion were prepared in advance for each session by the workshop planning committee. A member of the planning committee served as moderator for each session. Descriptions of each session and the guiding questions were provided to workshop participants prior to the workshop, and are provided in Appendix D of this document. Workshop participants regrouped for summary discussions at the end of each session. Discussions in these sessions were generally more abstract and conducted at a higher level than in Session 1. This level of conceptualization is reflected in the remainder of the summary.

## SNA FOR IMPROVING COMMUNICATION

Session 2a participants (see workshop agenda, Box 1-2) identified three inter-related long-term goals for improving communication for disaster preparedness and intervention:

1. The development of a networking planning tool, such as desktop computer software, to support homeland security. The tool could be designed to be used by an individual with at least some college education to define and model a given community. Such a tool would allow the user to identify community members, the resources available to the community, and the best pathways to disseminate information to the community to achieve a desired outcome;
2. A better understanding of networking theory. This includes processes by which individuals emerge as leaders within a network, and what happens to information disseminated across a network. With a better understanding of network theory, knowledge of various processes can be leveraged to support community resilience, create more effective support for communities, and to mobilize resources quickly when necessary; and
3. The means to conduct impact analysis (e.g., what happens in a network when specific information is disseminated) and scenario analysis (e.g., which of multiple scenarios will have the most desired impact).

Tools are currently available to accomplish some of these objectives, and new technologies are rapidly evolving. However, many existing tools may be too academically oriented for practical application or are not packaged for the specific needs of the disaster management community. The most effective user interfaces for computer programs could be developed with the full cooperation of emergency management practitioners and an understanding of the practitioners' needs. Workshop participants noted that to get the most out of networking tools, practitioners will need training in their use. Interpretation of network visualization graphs (Figure 2-1, for example) will need to be incorporated into the training of law enforcement and first responders. Training in the use of networking tools would also be essential at the community level to educate the public on topics such as how to send text messages, and how to communicate with family members during emergency events.

The next three sections of this document discuss in greater detail the three objectives stated above.

### A Network Planning Tool for Practitioners

Many workshop participants stated that a network planning and visualization tool for the emergency management practitioner cannot be designed to perfectly plan for every disaster or reach every individual in a community. However, such a tool could be invaluable in helping managers maximize connectivity within networks and building community disaster resilience.

Building the tool could be accomplished in phases. Initial phases could allow basic SNA functions as described in Chapter 2. Emerging technologies, such as data mining and different analysis techniques, could be incorporated into future phases. The goal of the primary phase could be to provide planning support for the Department of Homeland Security (DHS), community leaders, and other community groups. Initial challenges will likely be indentifying the capabilities the tool should have, identifying aspects of networks that need to be targeted, getting data into the tool, and developing the capacity to maintain and update the data. Workshop participants stated that for planning purposes, the tools would give practitioners an understanding of a community's adaptive capacities during different emergencies. Sufficient flexibility to accommodate dynamic networks and technologies would make the tool more useful.

Workshop participants considered lessons learned from organization theory research. Optimizing a network could reduce the adaptability of that network; for example, optimizing a network to increase the speed with which goods are delivered during nonemergencies may involve choosing specific resources and transportation modes. Without backup plans and flexibility during times of crisis, people may be put at risk if the goods are unavailable from regular sources, or infrastructure fails. If the goal is to increase resiliency, it is essential that the network's adaptive capacities—its abilities to function under stress—are encouraged to expand and remain flexible. Given this information, workshop participants discussed incorporating network optimization functions into planning and visualization tools only with proper caution so as not to limit flexibility.

## The Theory of Networking

A better understanding of network theory can lead to better knowledge of communication processes. This kind of knowledge can be leveraged to support community disaster resistance, resilience, and recovery. The practitioner may be better informed on how networking tools such as Twitter and Facebook can be used to advantage. Understanding how networks and networking tools function is vital for practitioners, but it is also essential that practitioners understand how to create and distribute a robust message so that the correct message is sent to appropriate networking sites and media. Marketing experts may provide insight regarding how to create and distribute correctly received messages. Theory has not yet been developed on how to accomplish this.

Participants indicated the importance of basic research on concepts related to trust and online networks. Community members have different levels of trust in data sources and in the digital networking tools through which data are transmitted. For example, many individuals may be distrustful of information coming from community leaders or from networking sites community leaders support. Understanding issues of trust can help practitioners more effectively use social networks to convey their messages. Understanding the processes by which individuals or groups of network members gain trust within their networks to become online opinion leaders is another area to be explored. Practitioners may be able to take advantage of the same processes and gain the trust of networks, or they may be able to predict who the opinion leaders will be and meaningfully engage with these potential leaders to encourage the spread of important information.

On the infrastructure side of the issue, the means of managing bandwidths required for social networking is challenging. Understanding the impact of the use of various cyber-enabled communication tools during a disaster situation is a new endeavor, and difficulties encountered when encouraging the public to use the communication channels designated by emergency managers is exacerbated during disaster situations. Theory building and data analyses in these areas would also be beneficial.

## Impact and Scenario Analyses

The third of the long-term communication goals described earlier is the development of tools to conduct impact and scenario analyses. The tools could include those that (1) explore the best ways to disseminate negative news without making a bad situation worse (for example, what happens in a network if a specific warning is sent to a specific media outlet); and (2) choose the best of several options given a specific scenario (for example, what happens if electricity is restored first to one location rather than another). Tools that allow the user to understand the impact of information and actions were considered useful by workshop participants, but their use is largely dependent on an understanding of networking theory.

Workshop participants noted that the need for tools for impact and scenario analyses is not unique to disaster management. Commercial tools are under development for marketing purposes, and the Department of Defense is developing tools for security purposes. Multiple private sector organizations are exploring tracking capabilities using text mining and text analysis techniques. Quickly and visually tracking the rapid changes that occur during an emergency is challenging, especially when changes are monitored over a variety of media.

## Challenges in the Use of SNA and Networks

Workshop participants observed that quality data are required to conduct SNA analysis, and that such data are often unavailable. The lack of data and the capacity to manage large datasets are impediments to the use of SNA for real-time applications. In some cases, data exist to populate SNA tools, but there may be legal issues regarding the use of private information by public entities, or reluctance among jurisdictions and organizations to share data. The use of networking tools, as noted by workshop participants, is also somewhat incompatible with the DHS National Incident Management System (NIMS) guidelines for managing domestic incidents.[1]

Quality baseline data were also considered essential to effective SNA. The current state of a community has to be understood before effective interventions can be optimally designed and implemented, and their impacts measured. However, some workshop participants stated that the cost of collecting baseline data through interviews, fieldwork, and other means is prohibitive.

---

[1]The DHS issued and revised the National Incident Management System (NIMS) to serve as a template to manage incidents of any size regardless of the cause, location, or complexity. See www.fema.gov/emergency/nims/index.shtm (accessed April 20, 2009).

The cost of developing and maintaining new technologies is also high. Champions of networking technologies within communities and in Washington, D.C. (where they could interact with the federal government or appropriate interest groups) could be identified and asked to encourage the development and use of network planning and analysis tools. The most effective champions would be able to communicate the utility of networking and analysis tools to people such as first responders and those empowered within the community to overcome political obstacles. Partnering with groups that are already developing tools for impact and scenario analyses may be an effective means of advancing their development and use for emergency management purposes. Including emergency management practitioners in all stages of the conversation regarding the promotion of networking technologies would likely yield the most promising results.

## SNA FOR PLANNED INTERVENTIONS

Participants of Session 2b (see workshop agenda, Box 1-2) discussed three main issues:

1. The ways that knowledge of social networks is currently used to support disaster management and build resilience;
2. Additional research needed to help improve disaster management and community resilience, namely, additional SNA of formal disaster networks, community-based networks, and the intersection between the two; and
3. Funding strategies to support the required research.

### Integration of Networking and SNA into Emergency Management Policy and Practice

Workshop participants shared examples of how some practitioners and policy makers have at least a rudimentary understanding of social networks. Local communities in Southern California, for instance, have tapped into existing social networks to disseminate preparedness information. Lay health educators from the Latino community use their existing social networks to distribute information on earthquake preparedness and to share coping strategies across communities. American Red Cross staff are seeking to build organizational linkages with business partners, creating a more robust network of people able to staff emergency shelters. SNA is being incorporated to some degree in models to measure the effectiveness of epidemic disease containment measures in the case of pandemic influenza.

Practitioners and researchers alike raised numerous questions regarding the integration of social networking and SNA into management policy and practice. According to some workshop participants, the research literature already answers some of these questions, but primary research is still needed on a variety of topics and scales related to disaster management agency responsibilities, community networks, and the interactions between them. For example, does being prepared for disaster within one network (e.g., where an individual is employed) make an individual more prepared in another network (e.g., at home or within a religious community)? The effectiveness of using networks to promote resilience, disaster readiness, response, and recovery at different levels of society down to the household level could be studied. The identification and roles of faith- and community-based organizations within a community need to be explored. Comparative studies on the ways social networks affect levels of community resilience in diverse places such as Israel, the United Kingdom, and Japan would help build understanding of the network characteristics that are successful in this context.

Several questions were raised by workshop participants regarding the role of organizational culture (for example, fire, police, emergency medical services, and public health departments) in building social networks, in the sharing of information in the disaster context, and in building resilience in general. Research on the ideal network among these groups and the role of technology in facilitating it could provide the guidance needed to create the robust and flexible networks that are sustainable during a disaster. Research on networks that cut across districts or local, state, and federal levels of government and disaster management goals could lead to study of how these networks could be improved.

Emergency management practitioners would benefit from understanding the difference between information and resource networks, as well as the implications of those differences, so that they may effectively use networks to plan and implement interventions. Understanding how to identify and communicate with the correct people and organizations within different types of networks would also be useful. Research on the characteristics and behaviors of groups that are effective at organizing themselves for action around the hazards in their communities could inform practitioners about which community behaviors are advantageous and which are not. Similarly, research on how emergent groups use social networking technologies, for what purposes, and how these activities could be improved would also be important.

## A New Research Funding Strategy

Several workshop participants observed that the research infrastructure does not support the longitudinal research necessary to understand and apply SNA for building community resilience. A substantial infusion of research support would be essential to develop and maintain both longitudinal and rapid-response research. Workshop participants discussed the idea of creating regional collaboratives of local universities, agencies, and businesses. These collaboratives could be funded with local, state, and federal resources, and could serve as repositories for regional baseline data. Their existence could encourage thorough baseline expertise on regional social networks and adaptive capacities, and they could be information resources for federal and local response agencies during times of crisis. The regions could be consistent with the 10 regions into which the United States is divided by FEMA. Both longitudinal and rapid-response research could be conducted within the framework of the collaboratives.

## SNA FOR ENHANCING IMPROVISATIONAL RESPONSE WITHIN NETWORKS OF ORGANIZATIONS

In thinking about how organizations work together during a disaster, the participants of Session 3a (see workshop agenda, Box 1-2) considered improvisation an important aspect of disaster response. SNA has the potential to reveal new ways to coordinate or influence the convergence of people, resources, and information to improve improvisational response—those activities planned in immediate response to changed conditions or resources. Different organizations (e.g., police, public utilities) are responsible for different aspects of response, and all benefit from knowing what the others are doing. An organization or individual that responds in unexpected ways outside the organizational mission context can be like a musician playing a sour note; discord can be the result. Questions arise regarding how to involve individuals and groups in the response process; who is responsible for doing so; and how is information about their involvement shared among networks. The data essential for making these decisions are often unavailable.

The next sections of this chapter summarize workshop discussion of how networks of networks may be used to foster improvisational response, communication, and resilience.

### Tools for Fostering Sustainable Ties

SNA is a useful tool for understanding the nonlinear nature of many ties within networks and organizations. It may also help identify how flexibility could be built into networks and organizations to allow for effective planning given uncertain circumstances. Practitioners are most successful working with teams that continue to function sustainably during the unexpected. Of importance to the practitioner is how to create ties with those groups with interests relevant to a particular problem. For example, developing ties with animal rights groups may be beneficial when dealing with problems associated with the increased number of stray animals following a disaster. Better ties with a large number of organizations will likely yield greater network resiliency during times of disaster. A greater probability of successful and timelier response and recovery may also result.

Baseline data help researchers understand the conditions necessary for building successful relationships to achieve desired outcomes. Because disasters are not contained within jurisdictional or geographical boundaries, building ties and brokering information across agencies and jurisdictions could prove effective. The knowledge base is increased and additional resilience is built into the combined networks. Baseline data and SNA may also help emergency practitioners determine the needed balance between efficiency and redundancy when developing relationships for resilient networks. Establishing redundancy in a network requires resources, but is essential in situations when a part of the network fails. Some workshop participants cautioned that redundancy that creates competiveness among network members should be avoided.

Sustainable ties can be built with interfaces between government and community-based networks. Traditionally, communication during a disaster has been one-way, from emergency management to the public. Many workshop participants predicted that two-way communication will be central to future disaster responses. The practitioner will be most effective by staying informed of the dynamic nature of relationships within a network and how these relationships may change day to day and in response to stress. This is essential to sustainable communication within the network. Technologies available today can already provide practitioners with many pieces of information that when combined, tell a story of what a network is doing and how it may be changing. Practitioners can use cell phones to monitor movement and receive status updates from the emergency management community or the public using networking tools, such as Twitter, to stay informed. SNA research conducted to determine how best to monitor constantly changing and emerging networks would aid practitioners. Avoiding infringement of privacy rights would be an aspect of this research.

## Chaos and Improvisation

Workshop participants discussed the critical need for baseline research to understand how a community and its networks function normally and under stress. Some functioning may be chaotic and emergency managers could benefit from understanding when chaos does not need to be controlled. Some chaos may appear in the form of ad hoc groups that spontaneously offer assistance. Effectively working with these groups without a plan or knowledge of their capabilities depends on the ability of practitioners to direct volunteer response efforts appropriately. Impromptu enthusiasm can be harmful if not controlled. For example, the delivery of unneeded supplies could create added burden for emergency managers and distract them from more essential responsibilities. SNA may help determine where volunteer group efforts can be used. Research on how to foster the ability to rely more on improvisational response could be useful. According to workshop participants, determining who will direct these groups is as important as how to direct them.

A census will never exist that lists all resources available during a disaster. Disaster management practitioners do not know what groups exist or who will ultimately be able to provide services once a disaster occurs. They may not have a full understanding of the critical dependencies for decision making (for example, who depends on whom and for what). Such gaps in knowledge limit the ability to improvise effectively. Unpredictable failure and re-emergence of networks add to the chaos. SNA research and the development of tools that help practitioners sort through this kind of chaos could be valuable.

Networks that emerge or re-emerge following a disaster may be unstable or loosely constructed. Understanding how these emerging networks are organized is more complex the bigger the disaster. Understanding how mistakes may occur during the re-emergence of networks, and of how to correct them, is essential if practitioners are to use the networks effectively in disaster response.

## Networks of Networks

The way SNA tools can be used to understand how networks function has already been discussed in this summary. SNA tools can be used beyond postdisaster case studies of individual networks. They could be used to understand how networks of networks emerge and evolve in a disaster situation, and may help identify the sustainable linkages among them. Given enough data, the commonalities and important characteristics that allowed these networks and linkages to survive can be discovered. However, accessing and using public data for the purpose of studying or constructing networks of networks is often challenging.

Knowing what public data exist and getting access to them are challenging issues that are inadequately resolved by the Freedom of Information Act (FOIA),[2] according to some workshop participants. Jurisdictions have independent protocols for recording data that may be incompatible with one another. As already mentioned, jurisdictions and agencies may be reluctant to share information. Some of the most useful data for analyzing and constructing networks could be the informal or confidential data representing personal communications between individuals and between different organizations. There are no means to obtain data from informal sources.

The ways the fire, emergency management systems, police, and public health organizational cultures help or hinder the process of building networks of networks for disaster management and resilience is not well understood. Understanding the optimal role of networking technologies in creating sustainable ties among the organizations could enhance communication and collaboration during normal operations as well as during emergencies. Because disasters do not proceed according to a plan, emergency managers may be able to use SNA and related technologies to help networks of organizations coordinate improvisational responses.

---

[2]The full text of the Freedom of Information Act (5 U.S.C. § 552, as amended by Public Law No. 104-231, 110 Stat. 3048) can be found at www.usdoj.gov/oip/foia_updates/Vol_XVII_4/page2.htm (accessed April 20, 2009).

4

# From Theory to Practice

Workshop participants discussed how properly targeted research in networking theory, the social and resiliency sciences, and in research translation--conducted in parallel with the development of Social Network Analysis (SNA) tools designed specifically for and with emergency management practitioners—could facilitate the adoption of SNA by the emergency management community. In the same way that the adoption of geographic information system (GIS) technologies has changed how decisions are made, the adoption of SNA has the potential to revolutionize the way in which organizations and communities function in general, and prepare and respond to disasters in particular. Workshop discussion was driven by the charge from the sponsor to identify current research and theories in the use of SNA for studying and constructing social networks. The preceding chapters of this document describe several information and technology gaps identified by workshop participants that potentially hinder the common application of SNA by practitioners. Consistent with the workshop charge, research areas that could fill those gaps were also identified.

Specific research needs identified by workshop participants are found throughout the preceding chapters. This chapter synthesizes several recurrent and overarching themes of the workshop discussions. As in earlier chapters, ideas presented here are those of individuals or groups of workshop participants and do not necessarily represent consensus among all those present, or the views of the workshop planning committee or the National Research Council.

## INCENTIVES FOR FOSTERING PREPAREDNESS

Some practitioners at the workshop were concerned that homeland security policies and funding structures often compel community leaders and emergency practitioners to function reactively rather than proactively. Planning is often done within the constraints of top-down policies that focus on protection and response rather than mitigation, and few incentives exist for communities to work toward resiliency. Disaster preparedness and continuity planning, although desirable, are often not well developed. Resiliency, a concept not well understood among emergency management practitioners, is not a part of the top-down emergency management culture. Policies are in place that can inhibit the use of innovative tools that may foster community resilience.

At present, disincentives to prepare for disaster may exist because a community may receive greater monetary benefits in the form of federal postdisaster aid. Although research findings indicate that communities can expect a four-to-one return for every dollar spent on disaster mitigation (Multihazard Mitigation Council, 2005), communities often do not take advantage of the expected savings. The return is not realized unless a disaster occurs and a reduction in recovery costs is observed. Additionally, mitigation planning may be thwarted by an inability to decide where mitigation is needed. A community may mitigate in the wrong way or be prepared for the wrong disaster. Under these circumstances, a community may suffer during a disaster and see no return on its investment.

Workshop participants have observed that at least some federal money is distributed as incentive for mitigation. For example, the Federal Emergency Management Agency (FEMA) instituted a pilot program to increase the federal costshare for communities who developed FEMA-approved debris management plans.[1] The program ended in 2008, but it reflected congressional interest in reducing the cost of disasters and rewarding communities for better disaster preparedness. Many participants expressed the view that incentive programs are helpful, but also expressed concern that they be carefully designed to avoid draining on resources.

Some participants noted that incentives can also come from the private sector, for example, through lower cost of insurance premiums for better construction or retrofitting practices. Property owners, contractors, and suppliers could see immediate financial benefits, and properties are better protected against disaster.

## UNDERSTANDING AND USING SOCIAL NETWORKS

A wide range of computer-based and other social networks thrive at many community levels. Determining how best to apply SNA to understand preparedness among these networks and improve preparedness at the household, community, and organizational levels is important. Understanding how to adapt preparedness plans and interventions to serve at-risk populations, such as those disenfranchised from community networks, is also important. Workshop participants noted that SNA could be used to understand how communities organize around hazards and how people and organizations use networks during disasters. The use of SNA for understanding the effective role of advocacy groups (for example, those supporting special-needs individuals) in building resilience and providing recovery assistance was also stressed. How social support and social embeddedness influence community resilience, and the importance of public connectivity in facilitating resilience are also topics meriting consideration.

SNA can be used to identify and study the characteristics and functions of trusted leaders, organizations, and information sources within networks. By understanding how leaders and organizations emerge within the computer-network environment under

[1]See http://www.fema.gov/news/newsrelease.fema?id=46906 (accessed May 4, 2009).

different circumstances, and understanding the characteristics that allow them to become trusted, emergency management practitioners could more effectively engage leaders and organizations in improving community resilience. Understanding the constraints under which decisions are made within organizations will also allow practitioners to target their communication with the organizations effectively. SNA could be applied to government organizations and networks, such as emergency response organizations, to study the attributes and functions that make them successful during all phases of a disaster. Once positive attributes are identified and understood, practitioners can build similar attributes into their own organizations and networks.

The social network community is a complex and interdisciplinary enterprise requiring the sharing of information in different ways among different members. Organization charts cannot explain the important relationships and transactions that take place between individuals and within and between organizations. SNA experts stated that better tools exist to display the same information and that practitioners could benefit from learning their use.

In general, those in authority benefit from the situational awareness of who is working with whom, on what tasks, and under what circumstances. SNA tools could help them visualize the important connections within networks. Effective disaster preparation is dependent on knowing before a disaster what groups and organizations work well with one another, which individuals know one another, and what sources of information are trusted. This is especially relevant if communication systems fail during an event. Workshop participants pointed out that optimal response depends on understanding what links remain active and which will quickly be restored following infrastructure failure. Critical links need to be identified and plans put in place so communication is sustainable during total infrastructure failure.

Communication is essential to situational awareness. Situational awareness gives practitioners an understanding of what is happening within the community to determine who needs what resources, and who needs to know what information. Workshop participants repeatedly discussed the need for practitioners to understand how to use their situational awareness to better disseminate the right information to the right people. SNA tools could be used to identify the right people and the most efficient means of communicating with them. SNA can then be applied to determine the impact of the information once sent. For example, in 2008, people on the Texas coast failed to evacuate despite warnings of the immanent arrival of Hurricane Ike. The perception among the public was that a category 2 hurricane was not dangerous. However, some workshop participants noted that had information of the storm's hazards been broken down into the dangers associated with wind, water, and storm surge, and had the information been conveyed effectively, people in the community may have reacted differently and the outcomes may have been less severe. SNA applied in this type of scenario could lead to better outcomes.

Study of how SNA or similar analyses are applied in areas such as network-centric warfare,[2] counter terrorism, and public health could be applied to SNA for improving community disaster resilience. The vocabulary of network-centric warfare is different

---

[2]Network-centric warfare is a Department of Defense doctrine based on using information technology to the military's advantage (Alberts et al., 1999). A well-networked military improves situational awareness and information sharing, resulting in the increased effectiveness of military missions.

from that used by social scientists during this workshop, but the goals are similar: to understand and improve how information is sought and exchanged, and to develop an action instrument that enables decision making. Understanding different patterns of network analysis and the different reasons and conditions under which the analyses are conducted could benefit both researchers and practitioners by helping them to characterize and use social networks for the advantage of communities. According to workshop participants, practitioners who are able to collect, analyze, understand, model, and incorporate network data into their decision-making processes will likely be better poised to help their communities become more resilient.

## QUANTIFYING ADAPTIVE CAPACITIES

According to some workshop participants, understanding and fostering community resilience implies the ability to measure a community's adaptive capacities—the skills and knowledge of a community that allow it to adapt to change—to different stressors over time. This is difficult given the dynamic nature of communities, especially during times of stress, and the external factors that impact them (e.g., changing legislation, availability of educational resources, changes in economic factors). Communities may require a variety of adaptive capacities to respond well to different stressors, and different communities may require different capacities to respond to the same stressors. It is important to quantify at what point a community is no longer able to function through a disaster—its capacity thresholds—and to understand the role social networks serve at different levels of functioning. An example was provided by workshop participants: the communities of Prince William Sound in Alaska were destroyed by a 1964 earthquake, and studies documented the physical, social, and self-help networks that formed in response. In 1989, the same communities were severely affected by the Exxon *Valdez* oil tanker spill. The networks did not come together following the 1989 event, and a social chaos was created that exists today.

More empirical data could help researchers understand how communities react to stress and determine under what conditions they might fail. Data could help researchers quantify the adaptive capacities that lead to community resilience.

## TRANSLATION FROM RESEARCH TO PRACTICE

According to workshop participants, scientific literature on SNA is fairly robust, and literature in disaster and community resilience is emergent. Until fairly recently, these research communities have had little interaction, and there has been even less communication between researchers and emergency management practitioners. Bringing multidisciplinary perspectives to bear on any given topic is difficult, but researchers and practitioners would both benefit from a greater exchange of information. Practitioners could learn state-of-the-art techniques that could be applied in their decision making. Researchers could obtain the data they need to refine their theories and models. Successful communication, however, is dependent on the ability of all parties to understand the terms related to social networks, SNA, and resiliency.

Workshop participants emphasized that translation research—research on how to enhance the adoption of research products into practice—is essential for learning how to move SNA from the research realm into application by practitioners in their communities. Transfer of technologies to practitioners is dependent on an understanding of which linkages among the research and practitioner communities need to be made. SNA could be applied to these communities to learn where the networks overlap and where relationships could be cultivated for better communication.

Researchers among the workshop participants stated that they and their colleagues do not know how to translate their findings for practitioners. Practitioners stated that translation would be most effective if findings were presented in a language and style that are applicable. For example, some suggested that to educate practitioners on the concept of resiliency, the paper written by Norris et al. (2008) on community resilience could be rewritten for journals such as the *Journal of Emergency Management*[3] and presented at conferences such as those of the International Association of Emergency Managers[4] and the National Emergency Management Association.[5] Just as important, researchers could involve practitioners in identifying research gaps and by creating the means to receive practitioner feedback on tentative research results. According to practitioners, receiving timely research results is important. Rapid response following an event is extremely helpful to practitioners, but it was recognized that not all research results could be shared quickly. A balance between rapid response and long-term analysis would be most useful. Some practitioners described what they called "[Hurricane] Katrina fatigue." They indicated that many practitioners may now be numb to useful findings still being reported after the 2005 event. Because practitioners are "burnt out" from reading Katrina-related reports, the reports have become less useful as a means of translating important messages or information about new technologies.

## THE NEED FOR A MEASURING STICK

Practitioners noted that from a government perspective, there is a need to establish intervention effectiveness measures before priorities can be set and resources allocated. Government agencies are more often required to show the impact of their activities and investments to validate expenditure of resources. For example, before a program is put in place to mitigate for a specific type of hazard, the means to measure how effective the program is once implemented have to be developed. To that end, participants stated it is essential to establish and quantify which adaptive capacities are most critical for building community resilience. Developing a "measuring stick" for social, economic, and relational capacities would be beneficial. However, because the connections among organizations are not fully understood, the status of the connections cannot be measured, nor can they be measured for change. Baseline data could provide metrics for change in networks, the characteristics that foster community resilience, and the magnitude of realized or potential stresses caused by a variety of stressors.

---

[3]See http://www.pnpco.com/pn06001.html (accessed May 8, 2009).
[4]See http://www.iaem.com/ (accessed May 8, 2009).
[5]See http://www.nemaweb.org/home.aspx (accessed May 8, 2009).

Workshop participants stated that with the application of SNA, practitioners could gain a better understanding of the thresholds at which communities fail, and could look at different thresholds of community fragility in response to different levels of stress. Empirical data on past disasters could allow estimates of failure thresholds for cities of different size and complexity, and with different types of networks among emergency response organizations. These kinds of post-event assessments of the effectiveness of social networks and the prevention of their failures are important components of baseline data. The data could allow the opportunity for researchers to document good practices and understand the characteristics of a community that enable resilience or rapid recovery. Participants also stated that comparative studies of networks that have emerged following a variety of disasters would be helpful.

Researchers do not have the capacity to collect much of the data needed to support SNA for building resilience, especially inter- and intraorganizational data. Many workshop participants stated that there is a great unevenness in the data available to populate databases, models, and SNA tools. As a result, comparative longitudinal studies of community resilience and change have not been conducted to any great extent. Because few resources are available to conduct much more than post-disaster case study research, the collective knowledge of the research community is focused on local systemic and episodic changes. The development of mechanisms for collecting network data, similar to surveillance mechanisms used for epidemiological monitoring, could be helpful. Additionally, workshop participants stated that data collection is underfunded, and that funding mechanisms do not exist solely for collecting and managing the large datasets required for SNA.

Baseline data as input to SNA can contribute to the understanding of how parts of a network draw on available resources during normal operations, and how those resources are stressed during a disaster. For example, baseline data on the normal operations of supply carriers and their distribution routes could allow the application of SNA to determine the best distribution options when service routes are disrupted or carriers become unavailable. Under stress, systems can fail catastrophically. SNA can help explain the extent of critical dependencies (who depends on whom) and the linkages between them. Analysis may allow planning of interventions to avoid systemic or cascading failures.

Workshop participants repeatedly stressed that the accuracy of network analysis, monitoring, and intervention design cannot be certain without baseline data.

## COMMUNICATION FOR RESILIENCE

Workshop participants stated that the ability to communicate as a vital capacity for resilience. Few things are more fundamental to community resilience than the ability to access, communicate, and use information. Information may be in the form of baseline data, media reports, and data that move across social networks. Development and transfer to emergency management practitioners of the correct technologies linking SNA to tools for collecting data, monitoring change, and conducting geospatial analyses are essential for reliable communication. With such tools, practitioners could potentially measure the quality of information received, determine what information is actionable, and determine what constitutes the best actions in timeframes that are useful during a disaster. Such tools could increase resilience by enabling two-way communication between emergency managers and the public.

By understanding how information is spread, and by understanding how trust is built between practitioners and the public and private sectors, practitioners may efficiently use networks to spread helpful messages and control rumors. Identifying the behavioral characteristics of those networks most effective at organizing themselves around hazards could help community leaders foster those characteristics in the networks within their own communities.

## NEXT STEPS

### DHS Call for Proposals

A purpose of this workshop was to provide the Department of Homeland Security (DHS) Human Factors Division of the Office of Science and Technology guidance on a potential research agenda to promote the use of SNA for improving community disaster resilience. This workshop was requested by DHS to inform a broad agency announcement (BAA) requesting white papers and proposals targeting research related to SNA and resilience. DHS plans to model awards after those given for the Small Business Innovation Research[6] program. In evaluating proposals, DHS considers metrics that show return on investments. Research outcomes may be products or knowledge. Workshop participants often stated that research partnerships between researchers and practitioners would improve networking between the two communities and reduce the need for translational efforts. According to DHS, the department cannot accommodate such partnerships through this BAA.

### Research Themes

Several research areas were identified by workshop participants as possible prerequisite to advancing the use of SNA for building community disaster resilience. Disaster management decision making depends on numerous factors, including the phase

[6]See http://www.dhs.gov/xres/grants/ (accessed May 8, 2009).

of the disaster, available resources, and the level of authority at which decisions are made. The application of SNA could improve the situational awareness of emergency management practitioners by allowing them to understand and measure the status of networks within their communities. Using knowledge gained through SNA, the necessary interventions and conditions and network associations required for their success can be identified. The best means of communicating and implementing interventions may be developed.

## Baseline Data

Baseline data include all manner of data regarding networks and their members. Without baseline data, researchers and practitioners do not know how communities and networks normally function; the impact of interventions may not be quantified; and the effectiveness of communication and operations may not be measured. Without a measure of community capacities to adapt to stress, interventions may not be effectively targeted for maximum benefit. Little is known, for example, about who populates the formal, governmental networks responsible for a region's disaster management or how they integrate with other social networks that reside in civil society. Without this baseline level of knowledge, it is difficult to evaluate how the compositions of social networks evolve and how this relates to resilience levels. Though workshop participants emphasized the importance of baseline data, current research funding makes the collection and management of baseline data not feasible.

## Validation Techniques

New networking technologies allow large amounts of data to travel quickly through networks. Workshop participants observed that mechanisms to validate new data, network models, and decisions made using SNA and related tools would be valuable to practitioners and scientists. Practitioners described needing mechanisms that vet for accuracy of data traveling through networks, and indicate whether information requires action or response. Practitioners need a means to categorize data as good, bad, redundant, and actionable.

## Understanding Network Dynamics

Building resiliency into social networks requires an understanding of the dynamic nature of networks and of how positive changes that prevent network failure during a disaster may be promoted. Because networks are constantly changing, workshop participants suggested that network models be constantly updated and that new methods for studying network dynamics are needed. Networks are likely to change quickly during a disaster and SNA tools could be more useful if they allowed the practitioner to quickly visualize the changing nature and uncertainties in relationships within and between networks. This would allow more effective diffusion of information at all stages of the disaster cycle.

### Better Data Gathering Techniques

New and more refined data-gathering techniques could result in better social network models. For example, workshop participants frequently stressed the importance of developing the means to obtain proprietary and personal data for SNA, while preserving the privacy of individuals and institutions. Data, such as who communicates with whom within and between private sector organizations and what kinds of people receive certain medical treatment under certain circumstances, provide important insights into the nature of networks and their members.

### Government and Community Interaction

How government networks interface with community-based networks for disaster management and building community resilience was often considered during the workshop. Workshop participants discussed how researchers and practitioners would benefit from greater understanding of the different ways that individuals, organizations, locales, and agencies are connected to social networks and understanding their use. An understanding of how their connectivity may change under stress would also be of benefit. Research on behaviors that emerge as a consequence of a disaster and research on the skill sets and attributes of individual and organizational network members that emerge as trusted opinion leaders could result in better identification of those able to collaborate with emergency management practitioners to build resilience.

### Exploring Community Resilience in other Contexts

Exploring other broader social issues where SNA has been or could be applied may provide useful information related to the application of SNA for building community resilience. Building resilience is not only about preparation for disasters. Research on how communities deal with issues such as ethnic oppression may yield a rich and pertinent literature regarding community resilience.

## New Research Paradigms

Although addressing these barriers was not directly part of the charge given the workshop planning committee, many participants stated that the barriers could affect the effectiveness of a future research agenda and slow the adoption of SNA tools by practitioners. Strategies to overcome these barriers were suggested by workshop participants and are summarized in the next paragraphs.

Workshop participants discussed how current strategies for funding research and its translation into practice are not adequate to address the large-scale and complex social science issues raised. New funding paradigms to accommodate larger and longer-term studies would benefit both the research and practice communities. One result could be better baseline data from which progress can be measured. Incentives to encourage multidisciplinary research and the rapid response efforts needed immediately following disasters could lead to results that are more immediately useful to practitioners. Workshop participants expressed the view that research collaborations among researchers and practitioners, and between public and private entities, could enhance the adoption of SNA techniques among practitioners. Issues associated with barriers of access to infrastructure and data could also be overcome. Collaboration between researchers and practitioners, with input from the private sector, could result in the most relevant research, tools, and data for decision making.

Among the workshop participants, some practitioners and researchers expressed concern that current homeland security priorities tend to encourage a focus on antiterrorism activities within the emergency management community. Some suggested that sources of community stress need to be adequately assessed to confirm whether a focus on antiterrorism is locally warranted. A better understanding of community stressors could allow for more informed allocation of resources.

Several workshop participants stated that researchers need incentives to collaborate with practitioners. Placing more value within the university and research cultures in research translation might foster such incentives. The medical community has begun to support translational research and activities, and has encouraged universities to consider the importance of translational work when making tenure decisions. Adoption of similar policies in other research communities could encourage younger researchers (those most likely to be familiar with social networking technologies) to do translational work.

The idea of developing regional collaboratives among local universities, agencies, and businesses was discussed among workshop participants. Local, state, and federal resources could be used to establish the collaboratives to encourage thorough baseline expertise on regional social networks and adaptive capacities. The regions could be consistent with the 10 regions into which the Federal Emergency Management Agency divides the United States.. Each of these collaboratives could serve as a repository for regional baseline data, and be a resource for federal and local response agencies during times of crisis. The longitudinal studies needed for disaster planning and rapid response investigations needed in the face of a disaster could tap those resources and be conducted within the collaborative framework.

## Broader Benefits of Community Resilience

It is relatively easy to determine how to mitigate damage to the physical infrastructure of a community. It is less straightforward to mitigate damage to the social fabric of the community and to make that community more resilient in the face of stress. Practitioners would benefit from understanding the critical interdependencies of a community using a "system of systems" approach. This could lead to better understanding of how disruptions can cause the cascading failure of infrastructure. It is essential to understand how to use emergent technologies to make strong and dynamic connections between people, information, and physical resources. Encouraging the creation of robust and flexible networks during times of normal operation could enable networks to be resilient during times of stress.

Many of the same capacities that allow a community to function during times of disaster (e.g., being well informed, well networked, and possessing the ability to respond to situations with creativity and flexibility) are those that allow a community to thrive during normal times. Many workshop participants expressed the view that by increasing the capacity for effective communication through social networks, a community may be created that is resilient to a broad range of stressors. Investing in building of community resilience is highly likely to yield rapid returns through the creation of stronger and healthier communities.

# References


Alberts, D. S., J. J. Gartska, and F. P. Stein. 1999. *Network Centric Warfare: Developing and Leveraging Information Superiority*, 2nd edition. Washington, D.C.: National Defense University Press, 287 pp.

Comfort, L. K., and C. Wukich. 2009. Designing Resilience for Communities at Risk: Building Capacity for Collective Action. In *Disaster Management: Global Challenges and LocalSolutions,* R. Shaw. and R. R. Krishnamurthy, eds. Hyderabad: Universities Press India.

Horrigan, J. 2008. Info on the go: Mobile access to data and information. Available at http://pewresearch.org/pubs/753/mobile-access-data-information accessed (April 7, 2009).

Latkin, C. A., D. Donnell, D. Metzger, S. Sherman, A. Aramrattna, A. Davis-Vogel, V. Minh Quan, S. Gandham, T. Vongchak, T. Perdue, and D. D. Celentano. 2009. The efficacy of a network intervention to reduce HIV risk behaviors among drug users and risk partners in Chiang Mai, Thailand and Philadelphia, USA. *Social Science and Medicine* 68(4):740-748.

Li, C., and J. Bernoff. 2008. *Groundswell: Winning in a World Transformed by Social Technologies.* Boston: Forrester Research, 282 pp.

Mackay, C. 1841. *Extraordinary Popular Delusions and the Madness of Crowds.* New York: Three Rivers Press, 740 pp.

Madden, M., S. Fox, A. Smith, and J. Vitak. 2007. *Digital Footprints: Online Identity Management and Search in the Age of Transparency.* Washington, D.C.: Pew Internet and American Life Project.

Multihazard Mitigation Council. 2005. *Natural Hazard Mitigation Saves: An Independent Study to Assess the Future Savings from Mitigation Activities, Volume 1 – Findings, Conclusions, and Recommendations*. Washington, D.C.: National Institute of Building Sciences, 19 pp.

Norris, F. H., S. P. Stevens, B. Pfefferbaum, K. F. Wyche, and R. L. Pfefferbaum. 2008. Community resilience as a metaphor: Theory, set of capacities, and strategy for disaster readiness. *American Journal of Community Psychology* 41(1-2):127-150.

NRC, 2007. *Successful Response Starts with a Map: Improving Geospatial Support for Disaster Management*. Washington, D.C.: National Academy Press. 198 pp.

Surowiecki, J. 2004. *The Wisdom of Crowds.* New York: Random House, 320 pp.

Wyld, D. C. 2007. The blogging revolution: Government in the age of Web 2.0. Available at http://www.businessofgovernment.org/pdfs/WyldReportBlog.pdf (accessed June 23, 2009).

# Appendix A

## Committee Biographies

**Susan L. Cutter**, *Chair,* is a Carolina Distinguished Professor of Geography at the University of South Carolina, and director of the university's Hazards and Vulnerability Research Institute. Her primary research interests are in the area of vulnerability science—what makes people and the places where they live vulnerable to extreme events and how this vulnerability is measured, monitored, and assessed. She has authored or edited 12 books, and more than 100 peer-reviewed articles and book chapters. Dr. Cutter has also led post event field studies of the role of geographic information technologies in rescue and relief operations in the September 11th World Trade Center attack and studies of evacuation behavior from Three Mile Island (1979), Hurricane Floyd (1999), and the Graniteville, South Carolina, train derailment and chlorine spill (2005). Most recently, in 2006, she led a Hurricane Katrina post event field team that examined the geographic extent of storm surge inundation along the Mississippi and Alabama coastlines and its relationship to the social vulnerability of communities. She has provided expert testimony to Congress on hazards and vulnerability and was a member of the U.S. Army Corps of Engineers Interagency Performance Evaluation Taskforce that evaluated the social impacts of the New Orleans and Southeast Louisiana Hurricane Protection System in response to Hurricane Katrina. Dr. Cutter serves on many national advisory boards and committees, including those of the National Research Council, American Association for the Advancement of Science, National Science Foundation, Natural Hazards Center, and the H. John Heinz III Center for Science, Economics, and the Environment. She is also a co principal investigator and member of the Executive Committee of the National Consortium for the Study of Terrorism and Responses to Terrorism, a Department of Homeland Security Center of Excellence focused on the social and behavioral sciences. She is a fellow of the American Association for the Advancement of Science and past president of the Association of American Geographers. Dr. Cutter is currently president of the Consortium of Social Science Associations. She received her B.A. from California State University, Hayward, and her M.A. and Ph.D. from the University of Chicago.

**Kathleen M. Carley** is a professor at Carnegie Mellon University's Institute of Software Research, Department of Engineering and Public Policy, and Department of Social and Decision Sciences, and director of the university's Center for Computational Analysis of Social and Organizational Systems. She specializes in research on organization theory, dynamic network analysis, social networks, multiagent systems, and computational social science. In her work, she examines how cognitive, social, and institutional factors affect individual, team, social, and policy outcomes. She is the author or coauthor of numerous books and articles in the areas of computational social and organizational science and dynamic network analysis. She served on the organizing committee for the National Research Council Workshop on Statistical Analysis of Networks and the Panel on Modeling Human Behavior and Command Decision Making: Representations for Military Simulations. She is a member of the Academy of Management, International Network for Social Network Analysis, American Sociological Society, American Association for the Advancement of Science, Informs, and Sigma Xi. In 2001, she received the lifetime achievement award from the sociology and computers section of the American Sociological Association. She is a founding and current editor of *Computational and Mathematical Organization Theory*. She received her Ph.D. in sociology from Harvard University.

**William A. V. Clark** (NAS) is a professor of geography and an adjunct professor of statistics at the University of California, Los Angeles. His research has focused on demographic change and the nature of the spatial outcomes of both internal and international population migration flows. He provided fundamental insights into the roles of preferences, discrimination, and public intervention in generating ethnic and racial segregation in America's urban mosaic. Dr. Clark has received several international and domestic awards, including the Association of American Geographers Award in 1987. He was elected to the American Academy of Arts and Sciences in 2003, and to the National Academy of Sciences in 2005. He currently serves on the NRC Geographical Sciences Committee. He received his B.A. and M.A. from the University of New Zealand, his Ph.D. in geography from the University of Illinois, and a D.Sc. from the University of Auckland.

**Eric Holdeman** is a principal in emergency management and homeland security at ICF International. His areas of expertise include building regional coalitions between agencies, governments, the private sector, and nonprofits. He has extensive experience in regional planning, emergency operations center design and construction, multimedia public education programs, joint information center formation and operations, media relations, and integration of technology into emergency management and homeland security programs. In 2007, he was recognized by *Government Technology* magazine as one of the top 25 people in the nation who "challenge convention, confront entrenched bureaucracy and promote innovation." Mr. Holdeman has authored numerous articles for professional journals and opinion pieces for local, regional, and national newspapers. Prior to joining ICF, Mr. Holdeman was the local emergency management director for King County, Washington, which encompasses the metropolitan Seattle area. In this position he established the King County Office of Emergency Management as a national leader in many areas of emergency management and homeland security. In 2005, King County was given a national award by the National Association of Counties for establishing a regional approach to homeland security. Additionally, the 9/11 Commission recognized the King County Regional Disaster Response Plan as a "best practice" for integrating the private business sector into community-wide disaster planning. Mr. Holdeman also served in the Washington State Division of Emergency Management for five years and in the U.S. Army for 20 years. He currently serves on the board of directors for the King County Crisis

Clinic and the Emergency Information Infrastructure Partnership and is on the advisory council for the Center for Regional Disaster Resilience. He is a past president of the Washington State Emergency Management Association. Mr. Holdeman received a B.A. in education from Concordia University.

**Randolph H. Rowel** is an associate professor at the Morgan State University School of Community Health and Policy and an associate faculty member at the Johns Hopkins University Center for Public Health Preparedness. Dr. Rowel's research agenda is to examine the cultural implications of public health emergency preparedness, response, and recovery. In partnership with the Maryland Department of Health and Mental Hygiene, Dr. Rowel recently completed a project that examined natural disaster experiences of low-income African American and Spanish-speaking Latino populations. He currently serves on the Maryland Bioterrorism Preparedness and Response Advisory Committee, and is director of the Why Culture Matters Work Group for Disaster Studies, an organization that informs public health professionals and faith and community organizations about the needs of vulnerable populations during natural and technological disasters. Dr. Rowel serves as an investigator for the Department of Homeland Security-funded National Center for Preparedness and Catastrophic Event Response, where he is working to establish the scientific foundation and principles of the practice of homeland security in matters of preparedness and response to catastrophic events. He received his undergraduate degree at Morgan State University and his master's and doctoral degrees from the University of Utah and the University of Maryland College Park, respectively.

**Monica Schoch-Spana** is a senior associate with the Center for Biosecurity at the University of Pittsburgh Medical Center (UPMC) and an assistant professor in the School of Medicine Division of Infectious Diseases. The Center for Biosecurity works to affect policy and practice in ways that lessen the illness, death, and civil disruption that would follow large epidemics, whether they occur naturally or result from the use of a biological weapon. Dr. Schoch-Spana, a medical anthropologist, has led research, education, and advocacy efforts to encourage greater consideration by authorities of the general public's capacity to confront bioattacks and epidemics constructively. She recently chaired the Center's Working Group on Citizen Engagement in Health Emergency Planning, and was the principal organizer for the 2006 U.S.-Canada summit on Disease, Disaster, and Democracy: The Public's Stake in Health Emergency Planning. In 2003, she organized the national summit, Leadership During Bioterrorism: The Public as an Asset, Not a Problem, and chaired the Working Group on "Governance Dilemmas" in Bioterrorism Response that issued consensus recommendations to mayors, governors, and top health officials nationwide in 2004. Dr. Schoch-Spana has served on the NRC Steering Committee of the Disasters Roundtable, the Committee on Educational Paradigms for Homeland Security, and the Committee on Standards and Policies for Decontaminating Public Facilities Affected by Exposure to Harmful Biological Agents: How Clean is Safe? She serves on the faculty for the National Consortium for the Study of Terrorism and Responses to Terrorism, a university center of excellence supported by the U.S. Department of Homeland Security. Dr. Schoch-Spana helped establish the Biosecurity Center of UPMC in 2003; prior to that she worked at the Johns Hopkins Center for Civilian Biodefense Strategies, starting in 1998. She received a Ph.D. in cultural anthropology from Johns Hopkins University and a B.A. from Bryn Mawr College.

# Appendix B

## Selected Recent References on Disaster Resilience, Social Networks, and Social Network Analysis

The first part of this appendix is a bibliography of references compiled in January 2009 by the National Research Council Committee on Applications of Social Network Analysis for Building Community Disaster Resilience. These references were provided to workshop participants in briefing materials distributed prior to the workshop. Given the broad range of research professionals and practitioners that were to attend the meeting, the committee prepared the bibliography in order to provide general background on the topics of social network analysis (SNA) and resiliency for those unfamiliar with the research or terminologies in these fields.

Workshop participants were invited to provide additional references. These are included in the final sections of this appendix.

### SELECTED BIBLIOGRAPHY PREPARED BY THE WORKSHOP PLANNING COMMITTEE

#### Books


Carrington, P. J., J. Scott, and S. Wasserman, eds. 2005. *Methods and Models in Social Network Analysis*. Cambridge, U.K.: Cambridge University Press, 328 pp.

Freeman, L. 2004. *The Development of Social Network Analysis: A Study in the Sociology of Science.* Vancouver: Empirical Press, 205 pp.

Grotberg, E. H., ed. 2003. *Resilience for Today: Gaining Strength from Adversity*. Westport, CT: Praeger, 281 pp.

Lindell, M. K., and R. W. Perry. 1992. *Behavioral Foundations of Community Emergency Planning*. Washington, D.C.: Hemisphere Press, 309 pp.

NRC. 2006. *Community Disaster Resilience: A Summary of the March 20, 2006, Workshop of the Disasters Roundtable*. Washington, D.C.: The National Academies Press, 15 pp.

Schoon, I. 2006. *Risk and Resilience: Adaptations in Changing Times*. Cambridge, U.K.: Cambridge University Press, 222 pp.

Scott, J. 2000. *Social Network Analysis: A Handbook*. Newberry Park, Calif.: SAGE Publications, 208 pp.

Tierney, K., M. K. Lindell, and R. W. Perry. 2001. *Facing the Unexpected: Disaster Preparedness and Response in the United States*. Washington, D.C.: Joseph Henry Press, 306 pp.

Wasserman, S., and K. Faust. 1994. *Social Network Analysis: Methods and Applications*. Cambridge, U.K.: Cambridge University Press, 825 pp.

## Articles

Anderson, W. A. 2008. Mobilization of the black community following Hurricane Katrina: From disaster assistance to advocacy of social change and equity. *International Journal of Mass Emergencies and Disasters* 26(3):198-218.

Beaudoin, C. E. 2007. News, social capital and health in the context of Katrina. *Journal of Health Care for the Poor and Underserved* 18(2):418-430.

Berke, P. R., and T. J. Campanella. 2006. Planning for post-disaster resiliency. *Annals of the American Academy of Political and Social Science* 604(1):192-207.

Black, K., and M. Lobo. 2008. A conceptual review of family resilience factors. *Journal of Family Nursing* 14(1):33-55.

Campanella, T. J. 2006. Urban resilience and the recovery of New Orleans. *Journal of the American Planning Association* 72(2):141-146.

Carley, K. M. 2006. A dynamic network approach to the assessment of terrorist groups and the impact of alternative courses of action. In *Visualising Network Information* (pp. KN1-1-KN1-10). Meeting Proceedings RTO-MP-IST-063, Keynote 1. Neuilly-sur-Seine, France: RTO. Available from: http://www.rto.nato.int/abstracts.asp.

Chen, A. C. C., V. M. Keith, K. J. Leong, C. Airriess, W. Li, K.-Y. Chung, and C.-C. Lee. 2007. Hurricane Katrina: prior trauma, poverty and health among Vietnamese-American survivors. *International Nursing Review* 54(4):324-331.

Colten, C. E., R. W. Kates, and S. Laska. 2008. Community Resilience: Lessons from New Orleans and Hurricane Katrina. CARRI Research Report #3. Available at http://www.resilientus.org/reports.html.

Cutter, S. L., L. Barnes, M. Berry, C. G. Burton, E. Evans, E. Tate, and J. Webb. 2008. Community and Regional Resilience: Perspectives from Hazards, Disasters and Emergency Management. CARRI Research Report #1. Available at http://www.resilientus.org/reports.html.

Cutter, S. L., L. Barnes, M. Berry, C. G. Burton, E. Evans, E. Tate, and J. Webb. 2008. A place-based model for understanding community resilience to natural disasters. *Global Environmental Change* 18(4):598-606.

Gunderson, L. 2008. Comparing Ecological and Human Community Resilience. CARRI Research Report #5. Available at http://www.resilientus.org/reports.html.

Harrald, J. R. 2009. Achieving agility in disaster management. *International Journal of Information Systems for Crisis Response Management* 1(1):1-11.

Israel, B. A. 1985. Social Networks and Social Support: Implications for Natural Helper and Community Level Interventions. *Health Education Quarterly* 12(1):65-80.

Joshi, P. T., and S. M. Lewin. 2004. Disaster, terrorism: Addressing the effects of traumatic events on children and their families is critical to long-term recovery and resilience. *Psychiatric Annals* 34(9):710-716.

Kautz, H., B. Selman, and M. Shah. 1997. Combining social networks and collaborative filtering. *Communications of the ACM* 40(3):63-65.

Manyena, S. B. 2006. The concept of resilience revisited. *Disasters* 30(4):434-450.

Mathbor, G. M. 2007. Enhancement of community preparedness for natural disasters-The role of social work in building social capital for sustainable disaster relief and management. *International Social Work* 50(3):357-369.

Moore, S., M. Daniel, L. Linnan, M. Campbell, S. Benedict, and A. Meier. 2004. After Hurricane Floyd passed: investigating the social determinants of disaster preparedness and recovery. *Family and Community Health* 27(3):204-217.

Morrow, B. H. 2008. Community Resilience: A Social Justice Perspective. CARRI Research Report #4. Available at http://www.resilientus.org/reports.html.

Norris, F. H., and S. P. Stevens. 2007. Community resilience and the principles of mass trauma intervention. *Psychiatry-Interpersonal and Biological Processes* 70(4):320-328.

Norris, F. H., S. P. Stevens, B. Pfefferbaum, K. F. Wyche, and R. L. Pfefferbaum. 2008. Community resilience as a metaphor: Theory, set of capacities, and strategy for disaster readiness. *American Journal of Community Psychology* 41(1-2):127-150.

Pais, J. E., and J. R. Elliott. 2008. Places as recovery machines: Vulnerability and neighborhood change after major hurricanes. *Social Forces* 86(4):1415-1453.

Ritchie, L. A., and D. A. Gill. 2007. Social capital theory as an integrating theoretical framework in technological disaster research. *Sociological Spectrum* 27(1):103-129.

Sadovich, J. 2008. Review-Disaster resilience: An integrated approach. *Journal of Homeland Security and Emergency Management* 4(4): Article 6.

Schoch-Spana, M., C. Franco, J. B. Nuzzo, and C. Usenza. 2007. Community engagement: leadership tool for catastrophic health events. *Biosecurity and Bioterrorism : Biodefense Strategy, Practice, and Science* 5(1): 8-25.

Schoch-Spana, M., B. Courtney, C. Franco, A. Norwood, and J. B. Nuzzo. 2008. Community resilience roundtable on implementation of Homeland Security Presidential Directive 21 (HSPD-21). *Biosecurity and Bioterrorism: Biodefense Strategy, Practice, and Science* 6(3): 269-278.

Troy, D. A., A. Carson, J. Vanderbeek, and A. Hutton. 2008. Enhancing community-based disaster preparedness with information technology. *Disasters* 32(1):149-165.

Vogel, C., S. C. Moser, R. E. Kasperson, and G. D. Dabelko. 2007. Linking vulnerability, adaptation, and resilience science to practice: Pathways, players, and partnerships. *Global Environmental Change* 17(3-4):349-364.

Walsh, F. 2007. Traumatic loss and major disasters: Strengthening family and community resilience. *Family Process* 46(2):207-227.

Zakour, M. J., and E. B. Harrell. 2003. Access to disaster services: Social work interventions for vulnerable populations. *Journal of Social Service Research* 30(2):27-54.

## ADDITIONAL MATERIAL SUGGESTED BY WORKSHOP PARTICIPANTS

### Books

Li, C., and J. Bernoff. 2008. *Groundswell: Winning in a World Transformed by Social Technologies.* Boston: Forrester Research, 282 pp.

Mackay, C. 1841. *Extraordinary Popular Delusions and the madness of Crowds.* New York: Three Rivers Press, 740pp.

Surowiecki, J. 2004. *The Wisdom of Crowds.* New York: Random House, Inc., 320 pp.

### Articles

Butts, C. T. 2008. A relational event framework for social action. *Sociological Methodology* 38(1):155-200.

Butts, C. T.; M. Petrescu-Prahova, and B. R. Cross. 2007. Responder communication networks in the World Trade Center disaster: Implications for modeling of communication within emergency settings. *Journal of Mathematical Sociology* 31(2):121-147.

Comfort, L. K. 2002. Rethinking security: Organizational fragility in extreme events. *Public Administration Review* 62(s1):98-107.

Comfort, L. K. 2005. Risk, security and disaster management. *Annual Review of Political Science* 8 (June):335-356. First published online as a Review in Advance on Mar. 17, 2005.

Comfort, L. K. 2006. Cities at risk: Hurricane Katrina and the drowning of New Orleans. *Urban Affairs Review* 41(4):501-516.

Comfort, L. K. 2006. Communication, coherence, and collective action: The impact of Hurricane Katrina on communications infrastructure. *Public Works Management and Policy* 10(4):328-343.

Comfort, L. K. 2007. Asymmetric information processes in extreme events: The 26 December 2004 Sumatran earthquake and tsunami.In *Communicable Crises: Prevention, Response and Recovery in the Global Arena*, D. Gibbons, ed. pp. 135-165 Charlotte: Information Age Publishing.

Comfort, L. K. 2007. Crisis management in hindsight: Cognition, communication, coordination, and control. *Public Administration Review (*Special Issue, Administrative Failure in the Wake of Katrina) 67(s1):S188-S196.

Comfort, L. K., and N. Kapucu. 2006. Inter-organizational coordination in extreme events: The World Trade Center attacks, September 11, 2001. *Natural Hazards* 39(2):309-327.

Comfort, L. K. and C. Wukich. 2009. Designing resilience: Building community capacity for action. In *disaster management: Global Challenges and Local Solutions,* R. Shaw. and R. R. Krishnamurthy, eds.. Hyderabad: Universities Press India.

Comfort, L. K., K. Ko, and A. Zagorecki. 2004. Coordination in rapidly evolving systems: The role of information. *American Behavioral Scientist* 48(3): 295-313.

Cutter, S. L., L. Barnes, M. Berry, C. Burton, E. Evans, E. Tate, and J. Webb. 2008. A place-based model for understanding community resilience to natural disasters. *Global Environmental Change* 18(2008):598-606.

Kalashnikov, D. V., Y. Ma, S. Mehrotra, R. Hariharan, and C. T. Butts. 2006. Modeling and querying uncertain spatial information for situational awareness applications. In *Proceedings of the 14th Annual ACM International Symposium on Advances in Geographic Information Systems* (ACM SIGIR), pp. 131-138. New York: ACM Press.

Lenhart, A., and S. Fox. Twitter and status updating. Pew Internet & American Life Project. Available online at http://www.pewinternet.org/PPF/r/276/report_display.asp. Accessed February 20, 2009.

Lind, B. E., M. Tirado, C. T. Butts, and M. Petrescu-Prahova. 2008. Brokerage roles in disaster response: Organisational mediation in the wake of Hurricane Katrina. *International Journal of Emergency Management* 5(1-2):75-99.

Mehrotra, S., C. T. Butts, D. V. Kalashnikov, N. Venkatasubramanian, R. Rao, G. Chockalingam, R. Eguchi, B. Adams, and C. Huyck. 2003. Project RESCUE: Challenges in responding to the unexpected. *SPIE Journal of Electronic Imaging, Displays, and Medical Imaging* 5304(January):179-192.

Mehrotra, S., C. T. Butts, D. V. Kalashnikov, N. Venkatasubramanian, K. Altintas, R. Hariharan, H. Lee, Y. Ma, A. Myers, J. Wickramasuriya, R. Eguchi, and C. Huyck. 2004. CAMAS: A citizen awareness system for crisis mitigation. In *Proceedings of the 2004 ACM SIGMOD International Conference on Management of Data*, eds. G. Weikum, A.C. König, and S. DeBloch, pp. 955-956. New York: Association for Computing Machinery.

Petrescu-Prahova, M., and C. T. Butts. Forthcoming. Emergent coordinators in the World Trade Center disaster. *International Journal of Mass Emergencies and Disasters*.

# Appendix C

## Workshop Participants

**Lauren Alexander Augustine**
Disasters Roundtable
The National Academies

**Victor Asal**
Rockefeller College of Public Affairs and Policy
University at Albany
State University of New York

**Carter T. Butts**
Department of Sociology
Institute for Mathematical Behavioral Sciences
University of California, Irvine

**Michael Byrne**
Homeland and National Security Division
ICF International

**Kathleen M. Carley**
Institute for Software Research International
Carnegie Mellon University

**Jay Chunn**
National Center for Health Behavioral Change
Urban Medical Institute
Morgan State University

**William A. V. Clark**
Department of Geography
University of California, Los Angeles

**Louise K. Comfort**
Graduate School of Public and International Affairs
University of Pittsburgh

**Susan L. Cutter**
Department of Geography
University of South Carolina

**Darrell L. Darnell**
District of Columbia Homeland Security and Emergency Management Agency

**Michael Dunaway**
Directorate of Science and Technology
Department of Homeland Security

**Eric Edkin**
Board on Earth Sciences and Resources
The National Academies

**David Eisenman**
David Geffen School of Medicine
University of California, Los Angeles

**Jared Eno**
Board on Earth Sciences and Resources
The National Academies

**Scott Feld**
Department of Sociology
Purdue University

**Jeffrey Friedland**
Emergency Management/Homeland Security
St. Clair County, Michigan

**Craig Fugate**
Florida Division of Emergency Management

**Jack Harrald**
Institute for Crisis, Disaster and Risk Management
George Washington University

**Arleen Alice Hill**
Department of Earth Sciences
The University of Memphis

**Eric Holdeman**
Homeland and National Security Division
ICF International

**Carl Latkin**
Department of Health, Behavior and Society
Johns Hopkins Bloomberg School of Public Health

**Richard Lempert**
Human Factors Behavior Sciences Division
Department of Homeland Security

**David L. Maack**
Office of Emergency Management
Racine County, Wisconsin

**Sammantha Magsino**
Board on Earth Sciences and Resources
The National Academies

**Matthew Morrison**
Pacific NorthWest Economic Region

**Fran H. Norris**
National Center for Disaster Mental Health Research
Dartmouth Medical School

**Jennifer B. Nuzzo**
Center for Biosecurity
University of Pittsburgh Medical Center

**Caetie Ofiesh**
Geographical Sciences Committee
The National Academies

**Randolph H. Rowel**
Department of Behavioral Health Sciences
Morgan State University

**Monica Schoch-Spana**
Center for Biosecurity
University of Pittsburgh Medical Center

**John Shea**
External Affairs
Federal Emergency Management Agency

**Ellis M. Stanley**
West Coast Emergency Management Operations
Dewberry

**Jeannette Sutton**
Natural Hazards Center
University of Colorado at Boulder

**Tricia Wachtendorf**
Disaster Research Center
University of Delaware

**Barry Wellman**
Department of Sociology
University of Toronto

# Appendix D

## Breakout Session Descriptions

The workshop agenda included breakout sessions during which workshop participants were divided into two groups. Participants of each concurrent session were to discuss specific issues related to themes defined by the workshop planning committee. The planning committee described the themes and provided questions to guide discussions. This section includes the text of materials regarding the concurrent breakout sessions provided to the participants prior to the workshop.

### CONCURRENT SESSIONS

#### SESSION 3: Improvisational Disaster Response and Social Network Analysis

This session examines opportunities for social network analysis in understanding and facilitating preparedness and response to emergencies in two different venues: networks among and between organizations and networks within local communities. In organizational response (Session 3a), we are interested in how more productive interactions and flexibility for improvisation and how these can be enhanced through social network analysis and social networking technology. For local networks within communities and among individuals (Session 3b), the concern is how to empower and engage such communities and foster collective behavior in preparing for, responding to, and recovering from disasters.

#### Session 3a: Improvisational Disaster Response and Social Network Analysis: Networks of Organizational Connections

This session examines opportunities for social network analysis in understanding and facilitating preparedness and response to emergencies in two different venues: networks among and between organizations and networks within local communities. In organizational response (Session 3a), we are interested in how more productive interactions and flexibility for improvisation and how these can be enhanced through social network analysis and social networking technology. Participants are provided the following questions to help guide the discussion. The session rapporteur will summarize the discussion to all workshop participants during the plenary session following the discussion.

1. How do social networks operate organization to organization and what are current knowledge gaps in our understanding of their functioning?
2. To what extent is this currently measured using Social Network Analysis (SNA)?
3. To what extent is SNA used and how can SNA facilitate, maximize, and foster pre-conditions that will permit improvisational response during disasters?
4. What SNA research or change in practice is required to facilitate more productive interaction between the Incident Command System (ICS) and the emergent networks and responders? What tools would be most useful in this effort?
5. How can social networking technology be used to maximize the utility of emergent responders?
6. How can SNA be used in choosing what social networking technologies should be utilized?
7. What are the gaps in knowledge and technology that inhibit the application of social network tools and theory to improvisational response?
8. What are the barriers in terms of training, technology, and policy to applying social network theory and tools to improvisational response?

**Session 3b: Improvisational Disaster Response and Social Network Analysis: Networks within Local Communities and Between Individuals**

This session examines opportunities for social network analysis in understanding and facilitating preparedness and response to emergencies in two different venues: networks among and between organizations and networks within local communities. For local networks within communities and among individuals (Session 3b), the concern is how to empower and engage such communities and foster collective behavior in preparing for, responding to, and recovering from disasters. Participants are provided the following questions to help guide the discussion. The session rapporteur will summarize the discussion for all workshop participants during the plenary session following the discussion.

1. What Social Network Analysis (SNA) tools are used currently to identify pre-existing networks in communities?
2. How is and can SNA be used to understand the collective and emergent behavior of social networks before, during, and after the disaster cycle
3. How can the collective and emergent behavior of social networks be fostered and directed before, during, and after the disaster cycle?
4. What technologies are most appropriate for engaging and empowering social networks at the community level?
5. What SNA tools can be used to determine what we don't know about how such social networks are built, function, grow, prosper during both pre- and post-disaster? What are the gaps in knowledge and technology that inhibit the application of social network tools and theory to improvisational response?
6. What are the barriers in terms of training, technology, and policy to applying social network theory and tools to improvisational response?